BORN WILD
JOURNEYS INTO THE WILD HEARTS
OF INDIA AND AFRICA

SWATI THIYAGARAJAN

B L O O M S B U R Y

NEW DELHI • LONDON • OXFORD • NEW YORK • SYDNEY

BLOOMSBURY INDIA
Bloomsbury Publishing India Pvt. Ltd
Second Floor, LSC Building No. 4, DDA Complex, Pocket C – 6 & 7,
Vasant Kunj, New Delhi 110070

BLOOMSBURY, BLOOMSBURY INDIA and the Diana logo are trademarks of
Bloomsbury Publishing Plc

First published 2017

ISBN: 978-93-90077-08-3

2 4 6 8 10 9 7 5 3 1

Printed and bound in India by Replika Press Pvt. Ltd

Bloomsbury Publishing Plc makes every effort to ensure that the
papers used in the manufacture of our books are natural, recyclable products made
from wood grown in well-managed forests. Our manufacturing processes conform
to the environmental
regulations of the country of origin

To find out more about our authors and books visit www.bloomsbury.com and
sign up for our newsletters

BORN WILD
JOURNEYS INTO THE WILD HEARTS
OF INDIA AND AFRICA

To,
Siddharth Butch,
or uncle Siddharth as I knew him.
Thanks to you I now know that there is a world,
'beyond my nose'

CONTENTS

FOREWORD

'IF A tree falls in a forest and no one is around to hear it, does it make a sound?' This is a book that tells the stories of sounds and sights that most of us never even get close to. It distills over 20 years of dedication to understanding and reporting Nature and the environment. This book may have been a long time coming – and, for me, it has been worth the wait.

Swati Thiyagarajan, now our Environment Editor, has been a pioneer in the field of environmental reporting and has steered most of NDTV's initiatives in raising awareness about key issues affecting us today and future generations. She has won numerous national and international awards for her reporting – such as the Carl Zeiss award for her reports on tiger conservation. She has also won the Ramnath Goenka award twice for India's best environmental journalist and her reporting has been instrumental in NDTV winning the 'Wind beneath my wings' award from Sanctuary Asia for *Born Wild* as well as the Indian television award for the best series on TV. Swati was also the first Indian journalist to be invited to be a judge at the prestigious Wild Screen festival, otherwise known as the Green Oscars.

Swati's strength lies not just in her enormous knowledge and experience of Nature – it is her ability to give it a greater context, to analyze it and relate it to the most pressing issues of our times. There is so much to enjoy and to learn in this book – from anecdotes and stories from the depths of the wilderness to fascinating backstories and the linkages with the bigger environmental crises that threaten our planet. It is a book about these linkages – and it is these 'linkages' that makes this book particularly special.

Born Wild was the first show of its kind on prime time on a news channel. Swati made it possible for NDTV to push the envelope when others thought of environmental reporting as esoteric and arcane. I remember the first episode of *Born Wild* on Olive Ridleys, the endangered turtles that come to nest in Orissa and die in the thousands as a result of illegal trawling. The impact and the response was huge and we decided – regardless of mass viewership levels – to continue to focus on similar topics – and Born Wild grew into a path-breaking series.

Swati has also helped NDTV launch some of our most popular campaigns such as the Greenathon; Save the Tiger; air pollution in our cities and many more. These projects – for many of which Swati was the content editor – signal where our priorities truly lie – with India's environmental and conservation challenges, and now the looming threat of climate change. As the first television channel to run such sustained campaigns over many years, NDTV is committed to continuing these regardless of whether market forces agree or not. Several of these campaigns have run live for 24 hours non-stop – something not done before by any other television channel – and much of it was possible because of Swati's insights.

This book revisits many of Swati's most memorable stories – spanning the length and breadth of India and southern Africa – brought to life over the years and now with fresh updates and new experiences. I strongly recommend you read *Born Wild* – it's an outstanding book from an outstanding environmental journalist.

February 2017   **Dr. Prannoy Roy**
New Delhi

ACKNOWLEDGEMENTS

THEY SAY it takes a village. In my case it took a small chunk of the universe to help me do what I do and write this book. This is the best part, to acknowledge everyone who played a part.

To my family. Dad, you always understood my slightly different drummer and loudly encouraged me to follow my dreams. Mom, through all the zillion 'wash your hands' admonishments, after I had petted yet another snake, dog, lizard, bird, cat, etc., you were baffled but equally supportive of my path. Krishna, thanks to the fact that you stayed the steady, committed one, looking after the family, I had the freedom to do what I wanted.

To Radhika and Prannoy Roy, my bosses, my 'other parents'. Often one is lucky to have a supportive family, then one grows up, leaves the safe world and faces the real world. You made the real world as nurturing, encouraging and loving as being back in my safe world. You made all of the opportunities possible.

To NDTV, my home away from home. How many people are lucky enough to say their first job is still their only job twenty years later? How many are even luckier to say they have loved every single day and have colleagues they respect as friends and family?

To Gargi Rawat. I could not have done *Born Wild* without you.

Mandakini Malla, Shiva, Pranav Dutt, who helped me put so many of my episodes together. My core team who made work fun every single time.

To Bittu Sehgal, who mentored a young reporter, believed in her, encouraged her and still patiently answers all of the questions.

To the Olsen family, Karen, Jurg, Bianca and Caitlin. You shared your beautiful animal family with me at Jukani.

To Nina Subramani. Much of what you will read were nascent ideas we discussed over so many midnight 'orange juice sessions'.

To Brinda Karat. My Mashi Bri, You made me look beyond just the wild, to the people. Made me understand the crucial distinction between lifestyle and livelihood choices.

To Craig Foster. Before you, I understood the facts of things, but now I see the heart of them.

Thank you, Neel Soni, Kalyan Verma, Dr Anish Andheria, Tom Foster and Alwyn Myburgh for your wonderful photographs.

To Ayesha Kagal, for great patience and pushing me to do this book and helping me for years with my writing.

Thank you to Bloomsbury for publishing the book.

But most of all, thank you to the Natural World, for being my greatest teacher, inspiration, place of peace and path to all my joy.

INTRODUCTION

*'The greatest threat to our planet is the belief that someone else
will save it.'*

– Robert Swan

LET ME start out by saying that this is not a book on natural history,
or wildlife biology or nature travels or investigative reporting or fluff
pieces on the cute and cuddly or about peoples' movements to save
nature. This is a book that has all of the above in bits and parts as seen
through my eyes, my travels, my work, my research and the wonderful
people and animals who guided me on my way. So in short order, it's
a book of the natural world in the way I see it and have reported on it
for NDTV, on my show *Born Wild*.

I have had the privilege to travel across India and then more
recently southern Africa and report from all of the places I have visited.
So when the idea for writing the book came up I thought to myself,
why not compare, mix and match and just write about everything that
has ever touched me, changed my view of the world, challenged pre-
conceived notions and just made me understand that no matter how
much I might venture into nature, nature venturing into me has had
the far greater effect. I thought to myself, can I share that? If I can, will
it inspire more people to venture out and allow nature in?

We are living in interesting times. No, not the politics, which for
me are more depressing than interesting, but interesting in that we have
a decade now, if that, to change the way we are at present consuming
the world around us. We are in the sixth mass extinction wave. By
2050 there will be more plastic in the world's oceans than fish. Climate
change is here no matter if we agree or disagree on whether it is human
induced. According to climate scientists, our carbon budget is near

exhaustion. That is all the carbon every single person in the world can use before we reach a tipping point. We will be at 9.7 billion people by the year 2050 and according to Professor Emeritus of Microbiology, Frank Fenner, who helped wipe out smallpox, the human race faces extinction within a 100 years due to all of the previously mentioned issues facing us.

In an exclusive interview with Sir David Attenborough, which I have included at the end of the book he said to me, 'How do we expect to grow infinitely in a finite world?'

Scientists have warned that unchecked environmental degradation, deforestation and biodiversity loss is also bringing new diseases into the world. Case in point, the Zika virus, Saars, Swine flu, virulent Ebola and others. Right now there is the danger of smallpox coming back as the Siberian tundra thaws and exposes bodies that died due to smallpox.

Years of watching predators, has taught me a few things. Predators are key to every ecosystem. They usually take out the sick, old, slow, the very young, the occasional pregnant mother and the distracted.

Predators actually bring life to a prey herd and although what they do is killing, they make the herd stronger for it because each time the strong, the alert, the fast and the intelligent survive, the best genes keep getting carried forward. In this way a herd gets better and better and healthier.

So an apex predator by simple virtue of how they eat, keep an entire ecosystem alive. Their food sources need food that is food to other species and so on and so forth. Carnivore is key to biodiversity and the health of an ecosystem.

So one would think the Homo Sapien would be key to the health of the eco system because after all right now we are the apex predators.

Homo sapiens started out as herbivores and quickly became omnivores. It is a fact that the more specialist the diet of the animal, the quicker it will go extinct. Wipe out the only food they eat and boom, animal gone. In fact new theories now say that Neanderthals might have gone extinct when Homo sapiens became dominant, not because we killed them off, but because their diets were much more restricted than ours. So some climate, planetary or extinction change put pressure on them and they vanished. So, it's a good thing we eat everything. Things should have worked out well.

However, things are a mess and it's not because of what we eat, but because of how we go about acquiring our food. We have become an omnivorous predator that multiplies like prey. Every self-respecting predator knows one thing, never breed so much that you outnumber prey. We are breaking all the rules of the natural world. We also target in a big way even the breeding populations of the prey—something non-human animal predators knew well enough to leave alone. In any species the key to their survival apart from availability of food and water is the health of their breeding populations.

All of this has weighed on me in the approach to this book. Are we aware of this? Do we understand the immediate consequences of our actions on the natural world? Do we really understand what this means for the future of our children on this planet? Are we really aware of how close to the edge we are?

Are we aware of the power we have to bring about change and reverse this course?

Sir David Attenborough, said to me, to make people care we must first show them what there is to care about, why we should care and then hope that will make people on their own aware of the questions asked above.

Take the fruit fly. Genetically speaking fruit flies and we are surprisingly alike. According to NASA, about 61% of known human disease genes have a recognizable match in the genetic code of fruit flies, and 50% of fly protein sequences have mammalian analogues.

Infact we are so alike that some scientists are planning to send fruit flies into space and study them to make life easier for astronauts.

So if we can have that much connection to a fruit fly, it's hard to hold oneself separate from anything in nature. It's in the holding of oneself as separate and superior that things started to become a mess.

But is it too late?

When I asked Sir David if he was not disillusioned by the destruction he sees around him after a lifetime dedicated to the natural world he asked me to think about what we would have lost if some people had not worked to preserve it. He said that when things looked bleak it was reason for renewing the work and not giving up.

So this is my attempt to take you into the wilderness a little bit. To show you what we have, what we will lose and what we can do. I chose *Born Wild* as the title to my show on NDTV and for this book, as to me Born Wild, is the essence of what we all are. All of us, Kingdom: Animalia, Phylum Chordata, Class: Mammalia, Order: Primates, Genus: Homo, Species: Sapiens.

We might be 'civilized' or rather 'domesticated', but we are still part of this wonderful fabric of the natural world around us.

1

TIGERS AND GORILLAS
FLAME OF THE FOREST & SOUL OF
THE MOUNTAINS

God made the cat to give man the pleasure of stroking a tiger.
— Francois Joseph Mery

I WALKED with four tigers in the dark green gold of a forest. The
ground was zebra bright with strips of light and leaf shadows danced

under the tread of my elephant. Birds called from trees while alarm calls resounded through the jungle… of langurs, peacocks and cheetal. After all, not one or two but four kings and queens of the Indian jungle were on the prowl. The air was also filled with soft chuffs from the tigers as they rubbed against each other and butted heads. The chuff is an extraordinary sound, a gentle exhalation of air through the lips with a soft 'pffth' sound. It's a tiger's way of saying 'hey, hello, how are you!'

It was not a dream, not a vision, but a moment. A moment of being immersed in the Now, being immersed in every second that boomed as my heart beat in my ear. Tigers are usually solitary animals. The only time they are in a group is when they are cubs with their mothers and, when as sub-adults they cling together for a few last moments until nature and instinct drive them apart. Males and females spend some time together when they pick mates and also hunt together when they do so. Tigers are formidable hunters, hunting primarily between dusk and dawn. They attack by stalking, chasing and pouncing in a burst of strength and kill the prey by snapping their necks with a crushing bite to the throat. Nature designed them to be solitary for most of their time because a pack of hunting tigers would be entirely unstoppable. Nature, as everything else, requires balance and evens the playing field for the prey by making the tigers solitary hunters.

These four siblings had perhaps a few days, maybe a few hours of togetherness left. They had already left their mother who would soon rear a new set of cubs. The spirit of the bonds will linger a little longer until each one stands magnificent, alone, powerful and unchallenged – the blood and bones of our biodiversity in India. For now, they were still young, playful and naughty. They pounced on each other, rolled, chased shadows and nuzzled each other's faces. The living fire of their bodies, clashing and rippling in that dense forest.

The Bengal tiger is the second largest of the existing tiger species, Siberian tigers being the largest. Males can grow up to 300 kilos while females can weigh about 200-220 kilos. I felt I could step off my elephant directly on to their backs. That is how huge they seemed to me.

I have had the privilege of personally getting to know both Royal Bengal and Siberian tigers. I have seen them grow from cubs to adults and I have stood with them, played with them and even given them a cuddle or two when they were young. I got to do this thanks to friends of mine who run a big cat park and rescue animals from bad breeders or poor zoos and give them lifetime care. Normally, no physical contact is allowed with the animals. However, as I was helping them with ideas of how to make their predator park functional and be a voice for conservation, I was privileged enough to be around the animals. With over eighteen years of experience with wild animals, I have a deep and abiding respect for their strength and their nature, which helped me deal with the often rambunctious cubs. What that did for my understanding of tigers is phenomenal.

I have experienced firsthand just how powerful the muscles under the coat are. I have seen just how impossibly soft yet tough their skin is. Pricked my fingers on those ridiculously large whiskers, each one like a quill, intensely sensitive and sharp along the ends. Rubbed the soft noses and knife-like teeth and claws. I have felt them chuff against my hair, against my mouth as they have butted their heads against mine and rubbed cheeks to say hello. Today, these tigers are over five years old and I don't have physical contact with them anymore but even after a six-month gap, if I go to their vast enclosures and call them, they come running, chuffing, muscles quivering and press their faces against the fence so I might give them a quick kiss. It taught me that they are not just species, but individuals. It taught me that they are not just cats, but Cats. It taught me that they could kill me in under seconds with a swipe of one paw. The wild can be captured, can be captive bred, even trained, but it can never ever be tamed.

In Bandhavgad that day with the four tigers, while I let magic drop its cloak around me, I just felt the joy of that moment, that second, the joy of the four siblings while they were still together and felt the future of hope, the hope that this beauty would always grace the Indian jungles.

So of course I have to start my book with tigers, don't I? I don't care if it is clichéd. It has been my life and my passion. My work started with them; the orange cats that lit my nights as a child. Sher Khan, not as the villain from the *Jungle Book* but the mighty invincible heroes of Corbett's stories. The cat that dragged me into love with all that is wild and wonderful in this world. If I had a totem animal and I would like to think that all of us do, it would be the tiger. My reporting career in

NDTV started with the tiger and after twenty years, it still has a lot to
do with the tiger.

But while the tiger was the catalyst for my path in life, her
introduction to me was through an extraordinary source.

Maybe the first memory I have of the wild is the keening call
of the Brahminy kite wheeling over head as I plodded behind my
uncle Siddharth in the Theosophical Society. He used to wake me at
6 in the morning and drag me out for a walk in nature. Siddharth
Butch, naturalist, ornithologist, my father's best friend and the single
greatest influence in my life. Today, if I stand labelled as a conservation
journalist, well it is all because of him. The Theosophical Society in
Chennai is a vast green space that runs along the Adyar river all the
way down to the beach. It was here on this beach that Annie Besant
discovered J Krishnamurthy, the great modern philospher. It was walks
on that beach that created my deep nature connection. I did not know
it then while playing in the surf that I would have the opportunity
to explore that connection deeply as an environment journalist later
in my life. The Brahminy kite screamed as it dived straight into
the surf, its caramel brown-red feathers flashing in contrast with its
snowy white head. He landed in the sand, having eaten something
small on the fly, and kept a beady eye on us. The old broken bridge
that used to stretch across the river in the place where it met the sea
beckoned us to sit and watch the sun rise. Bathed in the slow growing
gold of the sun I remember hearing uncle Siddharth say, 'This
is all that is sacred.' A firm atheist, he did not hold with traditions
and rituals but was one of the most deeply spiritual people I have
ever known.

Of course, as a five-year-old that morning I can't say I understood
his words, but as an adult every day in nature has made me hear them.

My favourite story of his was about the dominant tigress of
Mudhumalai. Mudhumalai is a tiger reserve in Tamil Nadu. On one of
his dozens of expeditions into Mudhumalai he spent three days tracking
the tigress. They saw evidence of her everywhere but did not see her.
On the third night, while sitting around a fire near the old forest rest
lodge exchanging various stories of the day, two of uncle Siddharth's
friends went pale and stared over his shoulder and asked him not to
move. Now when that happens at night when around a fire, I think no

one can resist turning around to look behind themselves. So he did. And there she was, the tigress sitting just beyond the undergrowth in the clearing, watching the men who had spent the day tracking her. 'The tracker became the tracked' he would say and laugh. 'There is a lesson there for you, young lady,' he told me. As a child, I did not understand what he meant but in the years since, having spent my own time in wilderness, I know and feel what he meant. In the act of tracking an animal one loves and knows, one is actually finding oneself. It alludes to the greater oneness of us all.

Uncle Siddharth introduced me to the Guindy National Park and the Snake Park in Chennai. The main reason we went there was for the snakes. There I met and saw another extraordinary naturalist who I will talk about later, who would again change the way I looked at conservation in the later years. But that first day I went to Guindy National Park Zoo. I was there because Siddharth uncle and my Dad wanted a word with the zookeeper. They were unhappy about some of the enclosures and the conditions of some of the animals there. If uncle Siddharth was my mentor, my dad was my partner in crime. His great love for nature and its mysteries were something I inherited in my genes I like to think.

It was there that I had my first glimpse of the tiger. Not in a wild place nor even in a nice place. It was through the grim bars of a small enclosure, nothing more than a glorified cage, with cement floors and a small pool for water. I stood mesmerised by his eyes. It was as if he reached deep inside me and touched a live wire because it felt like an electric shock. I did not want to go anywhere else and I did not want to see anything else. After that first visit, the zoo was my weekend haunt. Today as an adult I remember the deplorable conditions of the animals, the miniscule cages, the smell of rotting meat and filth. I understand why my father was reluctant to take me there and why he and uncle Siddharth always seemed to have tense words with the keepers. Then, as a five-year-old, all I saw were the tiger's eyes. Between this amazing captive tiger and the stories of the wild tigers I would hear from uncle Siddharth, I was hooked for life. He would ask me to close my eyes and imagine stepping back into time 150 years ago – 'Think... think, what did the jungles look like? Imagine the world Corbett lived in while he tracked the various man-eaters in his stories.'

A curious cub who came so close to the car that my ordinary lens felt like
a telephoto lens in Umredh Karahandla Tiger Reserve.
Photo credit: Swati Thiyagarajan

'Imagine', he would say in his booming voice, his eyes lit and animated as he gestured, 'there would have been tens of thousands of tigers in the Indian jungle. There would have been deer, wild boar, birds of all colours and shapes and snakes and butterflies and insects. Now open your eyes and look. See the roads, cars, houses, look at how dirty the sea is, why my fishermen tell me that they have caught nothing today, and where are the tigers? See Swati, it all disappeared, that lovely jungle with all those animals in the time it took for you to close your eyes and open them. All gone in the blink of an eye. That my girl is the truth. We have as humans been here for the blink of an eye, and see what we have done, what we have lost. It's up to your generation now. Open your eyes, Swati, keep them open, don't blink, there is a world out there beyond your nose.'

I might have not known then what I could do with my interests and passions but at least I thought I could be like uncle Siddharth when I grew up. I imagined myself rushing around travelling, taking photographs and saving animals in distress. I got half of it right. I am sure you can guess which half.

By the time 1997 rolled around, the year I joined NDTV, the magic of thinking about Project Tiger as a great success was beginning to fade. In the late sixties, with tiger numbers having dwindled to a few thousand from the tens of thousands largely due to hunting, habitat loss and the growth and spread of the human population in India, a lot of pressure was put on the government to step up and protect the tiger. Project Tiger was established in 1973 by the Government of India. The project, a conservation programme was established to ensure that tigers would be protected and that a viable population of Bengals would always be found in their natural habitats. The government set up a tiger protection force to combat poaching and paid villages to move out of tiger forests. Unfortunately too often force and intimidation was also used to move the people out and to this day there are several villages in and around tiger parks that have severe grievances with the forest department. The tiger is not just an iconic animal in terms of its appeal and charisma but in the reality of the Indian ecosystem, it is an umbrella species. As Dr George Schaller, the greatest conservation biologist said in one lecture, the key to the health of any ecosystem is the carnivore. The presence of the tiger allows for the great biodiversity of our wild systems. By protecting a tiger, we are in turn protecting her habitat, her prey base, the food base and every other animal that is dependent on that ecosystem and habitat.

Throw one thing in this fine network out of balance and we have a domino effect we can't stop. So when we say Save the Tiger, we are saying save the Indian forests, save our biodiversity, save our water security and our carbon sinks.

Although many conservationists, biologists and wildlife enthusiasts were bemoaning the fate of the tiger and evidence of extensive poaching was mounting, the government decided to bury their heads in the sand and pretend like denial was just a river in Egypt. As a young reporter, it was exceptionally frustrating, for I needed facts and interviews to corroborate my stories. Only the conservationists and biologists were willing to talk on camera. If the government did speak, it was merely to issue denials.

I was appalled at how vindictively they went after one of India's most eminent scientists, Dr Raghu Chundawat who, during his radio collar studies of tigers in the central Indian jungles of Panna, between the years of 2004 and 2006, warned the government that the tigers were being poached. Instead of paying attention to the fact that we had one of India's best tiger experts sounding a clear warning, the government decided to persecute and discredit him for daring to raise questions about the safety of tigers in Panna's forests. Three years later, it was clear that Panna had lost all of its tigers. The big question – could it have been prevented if the government had only listened to Dr Chundawat? None of the officers in charge at the time when Panna lost all of its tigers were held responsible. In fact, they were given transfers and promotions. The great Indian bureaucracy chugging along! It's this apathy really that we need to fight every day. There are excellent forest guards and officers in service who have dedicated their lives to protecting the wild but always seem to lose out to the top-heavy brass and government officials who bog down the system.

It was officially declared by the Wildlife Institute of India in 2009 that all of Panna's tigers had either died or had disappeared. This was five years after the first warnings were sounded and two years after it had become quite clear to most people that Panna had indeed lost all of its tigers just like Sariska, which was declared tiger less in 2005. Sariska which had been established as a wildlife reserve in 1955 was declared a tiger reserve in 1978. Situated in Alwar, Rajasthan the park is part of the Aravalli mountain range making it rich in minerals. Despite warnings and even a ban on commercial activity within the tiger reserve by the Supreme Court illegal mining has been a huge problem in the park. In 2004, around the same time that Dr Chundawat was calling warnings about the tigers in Panna, day visitors and conservationist were raising worries over the tigers in Sariska. None had been spotted in a while and even indirect proof like pug marks, scratches and marks on trees and scat was not being found. The Rajasthan forest department instead of heeding the warnings, took the stand that the tigers had migrated out of the park due to the monsoons and that they would be back shortly. The fact that they had never done so before and the fact that tigers do not migrate due to the monsoons, did not stop the government from clinging to this foolish theory. Both Project Tiger and the National Tiger Conservation Authority latched onto this like it was their holy grail. There had been 16 tigers in the park in the previous year. In 2005, Jay Mazoomdar a journalist with the Indian Express then, broke the story that there were no tigers left in Sariska. This caused a furore and forced the government to investigate the matter. The CBI were called in and Sariska was declared a Project Tiger reserve without any tigers. Finally poaching was blamed for the disappearance of the tigers. I still could not get anyone in the Rajasthan government to talk about it. It also seemed that poaching was the main reason why the tigers disappeared. When the natural habitat of a park is not degraded and its prey base is in healthy numbers and yet the apex carnivore vanishes, poaching is the usual reason.

This tendency by the government to not respond to questions on tiger conservation and their lack of transparency on the ground in what they were doing to protect tigers and a weak forest department seemed to be the main reason why Project Tiger was in such a shambles. Panna and Sariska were the great wake up calls, a culmination to years of warnings.

My interest when I started in NDTV was more a look at understanding what tigers were. What were their habits? How did they live? What were some of the systems they followed? Why was it that poaching took such a toll and why could we not afford to lose them? To be effective as a conservation journalist, I believe firmly that a mere love for animals or nature does not suffice. One must learn, study, understand and know the facts. Here too I got lucky. Bittu Sahgal, editor, *Sanctuary Asia* magazine and conservationist extraordinaire, took great interest in a young green reporter and really helped me on my journey.

But all the dedication in the world still runs smack up against the absolutely obdurate government in denial.

I have personally been blacklisted by the Simlipal tiger reserve for daring to suggest that they might not have 101 tigers as they claimed. Simlipal is one of the largest Project Tiger parks and a crucial green belt in a part of Orissa, that is quite drought prone. The forests are a catchment area for several rivers. In 2005, the state claimed that there were close to 200 tigers in Orissa and of the 200, 101 were in Simlipal. In 2009, I did a story stating that the numbers were super inflated and that if there were more than 20 tigers there I would be surprised. I did this story after consulting with several conservation and wildlife experts in both Orissa and the Wildlife Institute of India.

It was a logical look at habitat, prey base and reports of tiger sightings. The Orissa Wildlife Board, which had not met in years, called an emergency meeting to blacklist NDTV and call me a liar. I met the Principal Chief Conservator of Forests (PCCF) and told him that I would apologise in public if their numbers proved right and asked him if he would do the same if my estimates proved right.

The new national census eliminated the pug mark method of counting tigers and used instead a far more scientific method of analysing habitat, prey base and camera traps. And in 2010, the census stated that the state of Orissa did not have more than 32 tigers, if that, as the margin was extrapolated at 15-32 tigers as they could not get to certain other sanctuaries due to Maoist presence. The Orissa government immediately rejected the report and said that there were double those numbers in the state, and then did not explain why, even at double, those numbers were half of what they had claimed for years.

Orissa was not my first brush with this sort of arbitrary way of doing conservation. Nagarjuna Sagar Srisailam tiger reserve in Andhra Pradesh, the largest tiger reserve in the country, a catchment area for the Krishna river, a lifeline in the state, was another park that lived in denial. Not only were they dealing with Naxals who would actually intimidate the forest guards from going into the park but they even quoted tiger numbers that were far higher than the reality on the ground. I could not blame the forest guards for feeling intimidated because at the time, almost ten years ago, most of them were just daily wagers and had no formal training. They barely had shoes or any other equipment to help them patrol the tiger reserve and there were just too few of them. They were no match against men with guns, knives and bombs. I could understand why they were more concerned for their own lives than protecting the tiger. In 2004 when I visited Nagarjuna Sagar Srisailam tiger reserve with my colleague Gargi Rawat, we drove through a park with burnt buses, blown up forest watch towers and huts and huge burnt swathes. We were told that all of this damage was by the naxals. The next day, we were asked not to enter the park as the naxals had called for a bandh. The forest guards were too scared to take us into the forest.

It has been almost fifty years since the establishment of the Wildlife Protection Act and the creation of national parks, sanctuaries and reserves. Our forest guards still struggle for their salaries, struggle to get decent equipment and struggle even to get confirmed jobs in many cases. They are not in the same category as the army or even the police and yet, every day, we expect them to physically protect the forests. A recent report has stated that more forest personnel die in India than in any other country. They are victims of both poachers and the animals they protect.

They are protecting our greatest resources and national treasures and yet they are untrained and demoralised. I spent a day with one of them and was told how his son died because in the monsoon he had to walk miles in flood-waters just to try and get his child to a doctor. He had himself, in his late thirties, been a daily wager for over fifteen years with no job security and was expected to singlehandedly patrol over a twenty square km area on foot. Not only are the poachers well armed and organised, so is the timber mafia, so are the Naxals and Maoists who operate in several forested areas. Even the villagers who come into the forest to graze their cattle or collect firewood sometimes have a knife. What do we expect the guards to do? Calling someone a guard and then giving him a stick does nothing to help. Many of them have no love for wildlife or forests and are just merely doing a job, the only one available as they have never been motivated to understand what they are protecting, why they are protecting it and been praised for their hard work. Until we take this duty seriously, and unless we have trained personnel on the ground who have the same rights as the police if not soldiers in the army, we are going to fail in protecting our biodiversity.

While this was the ground reality, in the corridors of power, the government was chugging along in a state of denial, ignorance and ad hoc decision making.

It seemed like a real mystery to me why the government was just so obdurate and refused to deploy various scientists and field researchers. If they were suspicious of independent scientists and conservationists then they could have at least consulted with the Wildlife Institute of India trained biologists and experienced conservationists to protect the national animal. There are fantastic individuals, too many to name here, all of them doing extraordinary work in conservation, wildlife research and science, and in my opinion, we just do not pay enough

attention to them or give them the time, support and respect they
deserve. We knew that tiger numbers were at an all time abysmal low
and we were still dithering on policy. There are at present 48 project
tiger reserves in India. Six more have been proposed, of which four
have been approved but not declared. An average citizen in India
would have probably heard of ten.

Tiger numbers in parks were earlier arrived at by counting pug
marks. Pugmarks are the paw prints tigers leave on the ground when
they walk. This method is a very inefficient way of keeping count as,
firstly, pugmarks are more visible on muddy or even clear ground but
not in undergrowth. Secondly, there was no real expertise within the
department that allowed them to discern between the same tigers pug
mark looking a little different on a new surface. So every pug mark
was counted as a different tiger. Thirdly, pug mark casts can be used to
create false marks to inflate numbers and that was something the parks
certainly did at the time just to make tiger figures look good in their
registers. These inflated numbers stayed on the books as there were
no independent audits and successive forest chiefs did not change the
status quo even if they knew the truth because then they would have
had to explain the drop in numbers.

In the meantime, NDTV had done a succession of stories
on poaching, habitat loss, and tiger deaths from poisoning and
electrocutions by local people who were trying to protect their cattle.
There was a time when every other day it seemed we were chasing a
story of a dead or dying tiger. One could not get either the environment
minister or the chief ministers or the forest department on record to
admit to the truth of a looming disaster.

It was late in the summer of 2002 when a Bangalore-based wildlife
enthusiast went to Nagarhole National Park to pursue his favourite
pastime, photographing tigers. A large male tiger seemed to be resting
in the dry golden grass. But much to the horror of the photographer,
and then the rest of the world when NDTV played out the footage
later that day on the news, the large male tiger had a massive steel
trap clamped on his paw. The tiger stood up shaking with pain and
was trying hard to get the trap off his paw. These steel traps which
are cheap but lethal implements are usually used by poachers to snare
small animals who are immediately crushed and killed in the jaws of

the trap. The agony the tiger must have been in was unimaginable. Using the footage, the forest department then tracked down the tiger and tranquilised him. The damage was so extreme that a part of his foot had to be amputated. Now came the dilemma. They could not release him back into the wild with an amputated paw. A great male tiger held in captivity in his prime is a huge loss to an ecosystem. He has a territory that he protects, he has females with whom he mates and cubs he protects with his very presence. Take out a male tiger in his prime and many dominoes come crashing down. But his injury was severe. It meant that he would never be able to hunt effectively in the wild. A hard decision was taken and they decided to keep him in captivity. This is a form of torture. A wild tiger suddenly having to endure the bars of a cage. Imagine yourself today, a free person in a democracy, suddenly being thrown into prison for the rest of your life. Masti – named after the place where he was captured, Mastigudi – spent the rest of his life in captivity, fathering a few cubs in the Bannerghatta captive tiger sanctuary.

A few days after Masti's capture by the forest department, they found a group of Phardis in the forest. Phardis are a small indigenous group from Madhya Pradesh who are superb trackers and hunters. Marginalised, with no prospects of reliable incomes, homes or education, many Phardis are employed to poach tigers and other animals. They were caught with many traps, knives for skinning, and other implements. It was on meeting and talking to both Belinda Wright and Nitin Desai of the Wildlife Protection Society of India that I learnt how a magnificent tiger could be killed with just a few hundred rupees invested. Mr Desai and Belinda Wright told me how the poachers watch the movment of the animals. Most big cats have a routine and they rarely vary from it. The poachers, once they know the main routes the cats took for drinking water or moving around, would place thorns in and around that path. The animals would avoid the thorns and be forced to walk into the path on which the trap would be laid. The traps were hidden in case the occasional forest patrol came across it. The poachers would kill and skin the cat on the spot, sometimes stretching out the skin to dry high on the trees. Then they would take the skin and wrap it around the women under their saris as they knew that the women would never get searched. In this manner the skin could be transported to where it had to go. It was shocking how well planned it all sounded. Both Belinda Wright and Nitin Desai, have spent their lives tracking the illegal trade in wildlife, often posing as buyers at great risk to their own lives to catch poachers. They have disguised themselves as buyers, used hidden cameras, cultivated sources that have lead them to raids and done it all despite death threats against them. They have helped nab many poachers, have provided evidence against others and employ lawyers with strong knowledge of the wildlife laws to prosecute the poachers and yet often poachers go free or the case takes too many years and evidence and witnesses vanish. Despite those set backs they still fight to keep the tigers safe.

In my personal opinion, it would have been kinder to shoot Masti as opposed to keeping him in captivity. Masti died in 2014, a full twelve years after his capture. I will sadly have to say that he died in vain because none of his suffering translated into any appreciable improvement, whether on the part of the government or the forest department to help protect our wild tigers better. In fact, I would say Masti effectively died on that day he was captured in May 2002, because that was the day his glory and his contribution to India's biodiversity as a wild primal dominant male tiger came to an end. His loss is a real tragedy as has been the loss of every wild tiger before him and after him. The loss of a dominant male tiger is incalculable. First, we lose a tiger in his breeding prime. Second, we lose any cubs he has fathered, as any other tiger taking over his territory will kill the existing cubs in order to bring the females into heat so he can mate with them and father his own cubs. Nature has designed this to be so in order to always ensure that the best genes are carried forward and establish the survival of the species. She did not calculate for the unnatural deaths of dominant males, allowing for lesser males to take over without challenges.

I remember watching B2, the big dominant male of the range area in Bandhavgad. I watched him mark and stalk his territory, spray his area at regular intervals and sniff for evidence of other tigers, or females if they were in heat. I watched him rise on his powerful hind legs, stretch himself to his full magnificent height and lift his forepaws over his head to scratch deeply into a tree. He did this to leave his mark for any challenging tiger that might come along. Another male in B2's territory will spot these marks, he will rise on his own hind legs and try and scratch above these marks. If he can scratch higher than B2 then he knows he is bigger than B2 and can challenge him, but if he can't reach the scratch marks then he will know he is not strong enough to take on B2.

I sometimes wish we humans had such elegant ways in which to sort out our issues without always having to go to war.

All tigers have a tough life. Dominant males establish territories, the size of which is dependent on the availability of food, water and females. In healthy forests where prey base is abundant and water available, territories can be small. If prey and water is scarce, territories can be larger. Female tigers while not having to fight for their territory nearly as much as males, still have the hard task of giving birth to their young, raising the cubs, keeping them safe from transient male tigers or other dominant males and predators like leopards when the cubs are young. Every ecosystem has a carrying capacity, or a saturation point. This is a fine balance. When a forest reaches its saturation point for carnivores, in that the dominant males have carved out spaces for themselves, other males born in the area have to leave until they reach maturity and can challenge an existing dominant for his territory. These males are crucial in their own way for future tiger populations because they are the ones who help spread the tiger populations in larger areas and carry fresh genes from one population to another, thereby preventing inbreeding. In areas where there are contiguous tracks of forests that are well protected, this dispersal helps tiger numbers grow, like in many parts of the Western Ghats. In places where forests are broken up, degraded, where ecological corridors are either crisscrossed by development projects or human presence, these dispersing tigers have a problem.

Like Broken Tail. A male tiger whose tail was broken at the tip and hence his name. His story would become important enough to become a movie. I first saw him as a beautiful sub-adult in Ranthambore. Two years later, a train mowed him down as he attempted to make a crossing after he left the park looking for his own territory.

These dangers are common to all dispersing animals that leave national park boundaries and move to establish new territories, but in

A mother with her two cubs. Watching tigers walk directly towards you is probably one of the most glorious wildlife experiences.
Photo credit: Swati Thiyagarajan

several national parks in the country like in Bandipur, Mudhumalai, Gir, Rajaji, trains, highways with speeding trucks, tigers, deer and even elephants are often knocked down and killed. For years there have been demands for these roads and railway tracks within protected areas to be blocked or at least monitored, but that is yet to happen.

Nature also takes its toll. Tiger populations, according to Dr Karanth, a tiger expert and conservation zoologist in Karnataka, can lose or gain 20 per cent of their population in any given year. Issues like cub mortality, infighting, injuries and old age are part of this loss. But in places like Nagarhole, where Dr Karanth has spent over two decades and pioneered and established the scientific use of camera traps in population density studies of large mammals, tigers thrive. Within these parks, rich prey density and diversity and protection allow females to have several cubs and raise them to adulthood. This keeps populations stable and growing gradually.

Today, Dr Karanth says that in some areas populations are actually saturated, which means for the area and prey and water available, the maximum number of adult tigers has been reached. Male tigers usually move out of their fathers' territory once they become sub-adults in order to prevent mating with their immediate family members and also help spread genetic diversity by moving towards a different population of females. When areas get saturated, this natural dispersion becomes frequent and the numbers of tigers outside of protected areas increase. Dr Anish Andheria, another big cat expert, recently set camera traps to prove that there were almost 50 tigers outside the forests of the protected Tadoba tiger reserve in Maharashtra in the extended Chandrapur division, one of the most populated, fragmented and disturbed forest areas. This level of disturbance with a lack of prey base can lead to conflict with people. Right now there is a huge crisis building in Chandrapur. Farmers living around these forest patches have complained about their fields being raided by wild boars. Wild boars can cause a lot of damage and wipe out an entire field of crops. For years their pleas have gone unheeded and now the farmers are angry. The government instead of working with them and helping them with compensations and non lethal methods of protecting their fields, like with solar fencing, flashing light electric fences also powered by the sun, decided to declare the wild boar a vermin and had over 300 of them shot. Wild boars are a prey base for the tigers and the tigers have been attacking and killing cattle as their prey base has been dwindling. Now the forest department is mulling the translocation of tigers. Translocating big cats is generally a bad idea. Its stressful on the animal and usually they don't like being moved out of their territory. They try and find their way back causing more havoc along the way as they are now both stressed and scared as they are out of the familiar places. Between 2015 and 2016, there have been more than 1500 cattle killed and 29 people too have been killed. In the greater Vidharbha region there are several protected tigers parks like Tadoba, Andhari, Satpura, Bor, Pench and so on and the tigers outside of these protected reserves are spill over populations. Fragmentation of forest corridors that used to connect all these various sanctuaries has forced the animals into pockets instead of them being able to naturally move into sanctuaries like Tipeshwar and then on towards the Sahyadris, in the Western Ghats and other forests. Tiger dispersal is natural behavior and the cats have nowhere to go or have roads, dams, villages and mines in their paths.

In 2015, there were several incidents of tigers attacking and killing people in both north and south India. Dr Karanth advocates the shooting of these animals that have killed people. In principle, it is a sound piece of advice. He says that by killing the offending animal, people feel safer and will be less inclined towards anger which they will then show by burning forests or poisoning and killing other tigers. The problem is, how are we to identify exactly which tiger did the killing? It's not as if just spotting a tiger in the vicinity of the killings means *that* is the tiger that made the kill. In any given area, there could be one dominant male, several females and even some transitory males and sub-adults. The man-eater in Ooty was shot but the tiger terrorizing people in Pilibhit has not been found and has stopped attacking people.

While tiger landscapes in places like the Western Ghats, the central Indian landscape and the Terai arc of Corbett and Valmiki that lead into Nepal have healthy populations, places like Orissa and adjoining areas like Jharkhand and Chhattisgarh have tiger populations that are close to extirpated. Even in the Eastern Ghats, in the Northeast, a bio-world heritage site, tiger numbers have been dropping drastically. Much of it can be blamed on poaching, inadequate protection and development projects. This is the unholy trifecta at the end of the day.

With the panic button being pressed on tiger conservation, NDTV decided to start their Save Our Tigers campaign. The campaign aimed to discuss the plight of tigers in the country, explore solutions to the problem and also help on the ground by collecting money which could be used to equip forest guards and help step up protection in the parks. In 2008, after the first official census figures with the new methodology of counting tigers was announced, the number which was placed at 1400, shocked the nation and NDTV did a small campaign, where we raised awareness and collected over one lakh signatures with the help of Sanctuary Asia and handed it over to then Prime Minister Manmohan Singh. The campaign worked so well that in 2010, we joined up with Aircell and Sanctuary Asia and started our big Save our Tigers campaign. Since then we have helped keep up the pressure on the government to protect our tigers. Armed with the first national census numbers, we collected a list of demands we then presented to the chief ministers of all tiger states in the country and even managed to extract a promise from then Prime Minister, Mr Manmohan Singh, that tigers would be made important to the national agenda. Our second campaign which was more of the same in 2012 stood on the strength of the first campaign and we saw tiger numbers stand at 1708 tigers. In 2015 we worked on the fourth Save Our Tigers campaign and we pushed forward a greater agenda to secure a viable future for tigers in the country. At present tiger numbers according to the new census figures stand at over 2000 tigers, a 30% increase in tiger numbers since 2010. A clear sign that media clamour, especially the NDTV campaign along with the din raised by naturalists and conservationists and indeed the citizens of India had finally prompted the government to adopt certain important measures to protect the national animal.

Most conservationists will tell you that these numbers are a silly way to look at tiger conservation. They will tell you that undue importance is being paid to the numbers. As a conservationist and a media person, I agree with both sides. Numbers might have no impact on the real conservation concerns on the ground, namely habitat loss, poaching, development and human pressure, but it most certainly helps lay people grasp whether tigers are doing well or badly in the country and it engages a regular viewer in the debate. It also has a flip side, which we are now seeing, which is a certain smugness sets in making the wrong people, i.e. the politicians believe that all is well with tigers and start to push for more development in the same areas that are needed for future tiger growth. Right now the Environment Ministry is more engaged in easing business then protecting the environment.

With the relentless media pressure on both Panna and Sariska, the government decided to re-introduce tigers in both parks. The Sariska exercise was both badly thought out and executed. All three tigers were brought in from Ranthambore with no examination of any existing close genetic links between the tigers. This can cause breeding problems in the future. Any source population that will be the founding stones of an expanding population must be genetically diverse to ensure a measure of success. If inbreeding is an issue from the beginning, the population is already weak and then becomes susceptible to disease and birthing issues. It turned out that the first three relocated tigers all came from the same father. Right now Sariska claims 13 tigers, of which five are adults who were relocated there.

The male tiger, who was moved from Ranthambore was a breeding male. He had already fathered young cubs and his removal upset that balance and the cubs disappeared.

Sariska also had an extensively degraded habitat and several villages within the reserve which had still to be moved out. Even today eight years after the re-introduction, Sariska has not bounced back in a big way. In 2010, one of the tigresses was even poisoned in the park in a tragic man-animal conflict incident, clearly indicating that while tigers were relocated, no long standing changes had been made on the ground, threatening the future of the tigers in the park. In contrast, Panna has not only bounced back, it has thrived. All the forest officers who had been in charge during the debacle were moved out. It's another matter that not one of them was prosecuted for the shocking mismanagement of the park despite a CBI enquiry that found collusion between the forest department and the poachers.

In 2009, two tigresses were moved to Panna National Park from Bandhavgad and Kanha respectively. This effectively eliminated any possibility of them being related. A third tiger, a male was moved in from Pench National Park. All the tigers were just older than sub-adults and none of them were established tigers with territories. Even then the male tiger from Pench made a beeline back to where he was found, a journey of over 250 kilometres. Sheer instinct took him back. This directional bonding, or homing instinct as it is known, has been recorded in many relocated cats. Even people who have domestic cats as pets will tell you how, when they are moved from home A to B, they might try and make their way back to home A. It is purely instinct. He was followed, trapped and tranquilised and brought back to Pench. This time, he followed the scent of a female tigress and mated with her. To ensure that he would find the female, the forest department actually did something unusual. They procured some tigress urine from tigers in the Bhopal zoo and sprayed it all over the area where the male tiger was released. By summer of 2010, Panna had its first set of four cubs. The adult tigers were labelled T1, T2, T3 respectively. The cubs were labelled 111, 112, 113 and 114 as they were the cubs in the first litter of the first tigress released, T1. So 1 for their mother, 1 for their litter and 1-4 for the number of cubs. Since then several more cubs have been born and three more tigers have been introduced here.

All six adults have settled in well and the cubs who themselves have become adults have had cubs. Today Panna is home to over 25 tigers. This success story ought to be a huge beacon of hope. However, the government in all its wisdom is contemplating quite a shocking and environmentally questionable project. They want to link the Ken and Betwa rivers, that are fed in and flow through Panna in order to create irrigation projects for farmers along the area. This project promises to undo all the gains as the over flow threatens to drown acres of forest bifurcating the park.

On a blistery, unseasonably cold and wet day, I sat on the back of a domesticated Asian elephant with cold drops of rain sliding down my back. The jungle closed in around us and I had my eyes peeled as we were trying to locate a female tigress and her cubs. In Panna, all the adult tigers, the six introduced tigers and the four cubs who themselves are adults today are all radio collared. This enables the forest department to keep an eye on them 24×7. There are three teams of men who pull eight-hour shifts round the clock every single day, rain or shine, following the tigers by jeep, elephant or even on foot. Having lost all the tigers earlier in the park, they are taking no chances this time. The biggest driving force behind such dedication was the Field Director of Panna, Mr Murthy, who himself spends up to eighteen hours a day in the field. Whether this level of dedication remains feasible is another question because the team is exhausted. It had been five years. The results, however, had been spectacular. Panna, today, stands at 25 tigers. From 0-25 in five years is a huge achievement. It also proves that tigers are very robust animals with excellent birth rates. Give them a reasonably inviolate space with adequate food and water, and the populations will increase.

Suddenly there was great excitement from the elephant in front of me as the mahout had spotted the tigress. She was on the move and her cubs were moving through the undergrowth, following her. In a flash I saw the cubs rush across to join their mother as she started to climb a rocky hill. The elephant followed. The rain picked up and hail started to fall. We were on paths that barely fit the tigress, so it felt like a miracle when the elephant followed. There were times when we were a few metres from the tigress who was walking parallel to us. I had one moment of wondering if I might find myself sharing my houda with

an irate female tigress. Tigers don't need a running start to jump high.
For the tigress to clear the 12 odd feet it would take to land on top
of my elephant would have been effortless. But I wondered for just a
moment because in the eighteen years I have travelled tiger country,
if there is one thing I have learnt about wild animals, it is that they
really do not want to come too close to people. They are wary and very
tolerant. Usually, it's the human 'idiot' factor that leads to an attack.
If an animal is surprised, or feels protective towards its cubs, or is in a
hunting mode and a human happens to be in its path, then too there
can be a confrontation. This tigress had figured out very quickly that
we were no threat to her cubs and she was only more interested in
getting them to a safe spot in the storm. To me it was a very special
experience because the real success of any re-introduction programme
lies not just in the re-introduced animals adjusting to their new homes,
but in having a new generation born in that home having their own
cubs. Animals will not either be calm or reproduce if they feel stressed.
These cubs I was looking at would be the second generation born in
Panna and, truly, here lies the long term future of the magnificent wild
cat in India.

I am not advocating relocation and re-introduction as conservation
solutions. I have seen a lot of it in Africa, many have been successful
and many unsuccessful. Ideally, it would be best to protect and ensure
the animals don't die out and disappear in the first place. But if it
is going to be done, it's best to do it in a way that brings peaceful
success because one is stressing out the animals that are being moved
and putting them through a huge challenge. What a waste to not get
it right!

But following this tigress, I was struck by another experience and
wondered if in India we would ever explore it.

Four years ago, I found myself in a van driving through winding
roads in Rwanda towards the Virunga mountains to the Volcanoes
National Park. Scenes from the movie *Gorillas in the Mist* played
through my head. I was here to go walking with mountain gorillas.
Growing up, I had three heroes. They were the famous 'trimates of the
primates', better known as Diane Fossey of *Gorillas in the Mist* fame,
Jane Goodall, the chimpanzee lady and beloved wildlife conservationist
and Birute Galdikas, the orangutan lady. I was mesmerised by their

stories and accounts of life in the wilderness. Their struggles, the challenges, the sheer audacity of being women in a hugely male dominated field at the time they started out, in parts of the world that had never seen white people let alone a white woman, like when Diane Fossey got to the Virunga mountains in Rwanda. Or when Birute walked the wet rainforests of Borneo. The three women forever changed how we would look at these primates and thanks to them, all three primates still have a future on this planet. While Birute and Jane Goodall are still alive, Diane Fossey was murdered brutally by the poachers she fought against. Her legacy, however, lies on these mountain slopes, now home to more gorillas than there once were. Her grave lies in these mountains, buried next to her beloved male silver back gorilla Digit. He was the first gorilla to let her into his world and was killed by poachers.

When I was a child, these women fired my imagination and lit inside me this great need to do something, to help preserve our natural world.

Here at the Volcanoes National Park, the government with the help of the local communities has created a unique tourism opportunity. Gorillas live in family groups. In this park, thirteen family groups have been identified by the park to facilitate tourism. These family groups are monitored and followed by teams of rangers who keep a 24x7 eye on them. Gorillas are a huge temptation to poachers and that, apart from habitat loss, was the biggest reason for their population crash. Today, even with all this protection, there are less than a thousand mountain gorillas left in the world. Originally, the family groups were followed merely for protection and research purposes. The animals in these family groups have become slightly habituated to the presence of people, over the years. The Rwanda government then decided to allow tourists to go walking with these specific gorilla families. Trained guides are appointed to each group and they walk you up the mountains. I was in a group of ten people. They do not take more than ten-twelve people per group. You have to be relatively fit as sometimes it could be a four-hour climb into the mountains just for a glimpse of one gorilla family. Every group is only allowed to follow one family. The guide stays in touch with the monitoring team through wireless handsets. My guide made my group gather and first took half an hour to instruct

us on various do's and don'ts. This is something that is sorely lacking in
eco tourism in our parks in India. Often visitors to parks in India are
loud, obnoxious and least interested in anything but spotting a tiger.
They often carry plastic items and other things in and then they litter.
The guides don't feel like they have the right to really discipline bad
behaviour when they see it. In Africa, they will simply throw a badly
behaved person out and sometimes he or she might get banned for life
from that park.

Our guide first talked to us a little bit about the Virunga
mountains and the gorillas. Then he assessed if we were fit enough
to take the walk. He warned us that there was a small chance that
even after a whole day in the mountains we might not see any gorillas.
The monitors only watch and follow the animals they do not interfere
with the animals' routine. Gorilla groups walk and feed and rest in
intervals and they can move quite quickly and cover vast distances.
They also move through dense forest undergrowth and might not
always be anywhere near tourist paths. We were told that they were
susceptible to human diseases like colds, coughs and even TB, and we
were to maintain a minimum seven-metre distance from the gorillas
at all times. We were also asked to speak in whispers, make no sudden
moves and generally be calm. Sometimes, my guide said, baby gorillas
and youngsters could be curious enough to come up to touch people
and we were asked to step back if it happened and not to reach out and
touch them in return.

Then we set off. The mist clung to the bamboo fronds and young
green leaves of the trees and very little sunlight filtered through into
our path. A light drizzle sang a song as it fell onto the dense canopy.
Our guide spoke to us about the gorilla family we were going to see,
giving us an understanding of gorilla behaviour. Gorillas are found
in tight knit family groups, that function like harems, with one big
male silverback, females and babies. In mountain gorillas, sometimes
1/3 of the group can be adult males who are all closely related.
Usually, gorilla males live outside of the group and some mountain
gorilla males can form bachelor groups. When a big male silverback
gets old, sick or dies, a subordinate or a stronger male can take over
the group. He also explained to us that two types of gorillas were
found in Africa. The mountain gorillas, which we were going to see

and the lowland gorillas, the ones that are mostly found in captivity around the world. Both types are critically endangered. The mountain gorillas however are found in just three countries spanning four national parks, Rwanda, Uganda and the DRC or the Democratic Republic of Congo. They live in forests found at heights of 650 metres to 4,000 metres. They live for forty to fifty years, and with all the unrest in the Congo, their population has gone a long time without international monitoring. The lowlanders, as their name suggests, live more on the plains, riverine forest habitats, tropical and rain forest habitats and on the lower slopes of mountains. This difference in both altitude and space has marked effects on what they eat. This dietary difference also determines behaviour, home range sizes and group sizes. They too are hunted and poached. Their populations are affected by habitat loss and human encroachment. As they are susceptible to human diseases, viruses like the ebola virus have wiped out whole populations.

In the early 1900s there were hundreds of thousands of lowland gorillas in western Africa, and in just the last few decades their populations have crashed by 90% in some cases. The lowland gorillas are found closer to where humans live and work and this has been a massive contributing factor to why they are dying out. One of the biggest threats to gorillas is the material named coltan. We might have not heard of it but all of us certainly use it. Coltan or columbite-tantalite is a metallic ore that becomes heat-resistant powder when processed. It's a vital component in the capacitors that control the current flow in cell phone circuit boards. The tech boom driven by consumers like you and me drove the prices of coltan up from 65 dollars a kilo to 600 dollars a kilo at one time. Now it is at 100 dollars a kilo, still a huge sum for people in a poor country like the Congo, the only place where it is found in vast quantities. While mountain gorilla populations in the Congo have halved in this destruction, lowland gorilla populations have crashed by over 90 per cent.

Coltan mining in the Congo has financed the civil unrest that has seen the death of over 5 million people. All of this has slowly sparked coltan consumers to insist that the coltan comes from only legal sources. This is, however, difficult to enforce. For the average consumer on the street it is hard to make a difference, but finding 'gorilla safe' coltan, and manufacturers who at least claim that their coltan source is gorilla safe, might be a start.

As raindrops touched my eyelashes, I thought of my cell phone nestled in my pocket and felt very guilty. If I, with so much more exposure to wildlife and its issues was so unaware of coltan, how can other people who don't do this as a career know?

It seemed to me unethical that cell phone manufacturers would not be upfront about this. Then it made me think of all the products I might be using in my daily life that could be affecting the tiger the same way back home in India. How much do we as consumers really know? And if we are interested in finding out, how do we go about it? Should we not ask the government to disclose these connections? We know now that the palm oil that we import from Indonesia and Borneo is killing the orangutans by destroying their forests. Palm oil is something that is present in our shampoos, soaps, chocolates and many other things. As coltan is not a visible or even a listed component in our smart devices, the only thing we can do is as consumers put pressure on cell phone providers and to ask for gorilla safe coltan to be used. This would mean that only certain legitimate mining operations that are conducted responsibly will supply the coltan.

With all these thoughts running through my head, I nearly missed my first glimpse of the gorilla. We were lucky as it had taken us less than two hours to find our family group. They had just finished feeding and were relaxing in an open grassland. The big male silverback was sunning himself while the females and babies milled about. It was the most astonishing sight. For that first incredible moment, I even forgot to breathe... this was a childhood dream coming true. They looked like incredibly large stuffed toys come to life. I had a baby gorilla stuffed toy I carried with me everywhere for a long time, even to college in Delhi. Bubba as I called him was my lucky mascot.

Even if there is another group of tourists who have not found their family group, they are not called and invited to view this group. They have to stick to their path and their group. The babies played and lolled and ran around while their mothers and older brothers and sisters played along with them, if they were not taking a quick nap. Big male silverbacks can get to being almost 6 feet 4 inches tall when standing upright on their two feet. They only do this when they want to intimidate. They then stand up, puff out their chests and roar in this terrifying way while thumping on their inflated chests, booming like a drum. Usually this happens between males who are challenging each other for territory or females. They might also do so to anything that they feel threatened by. Even on all four feet, gorillas are mostly quadrupedal, an impressive size. The amazing thing, though, is just how gentle they actually are. Standing on the slopes of the extinct volcanoes, looking deeply into the silverback's eyes, I just felt peace and calm. It was like looking into a mirror and seeing the best of myself, the best part of who we are, that gentle peaceful acceptance of being a creature who was woven into the very fabric of his world, surrounded by family, with his design and purpose firm.

In Panna, seated on top of the elephant watching the tigress I thought, could we do this in India? Could we walk with tigers? Here the adults were already collared. They are already being monitored 24×7. There are already people on foot and on elephants and jeeps following them. Yes, there is a difference between following a herbivore as opposed to a carnivore. But having walked with cheetahs and lions I know that it's possible. On any given day in India, hundreds of people across the country, from daily wagers to forest guards, patrol tiger parks on foot. Except for the odd attack there has never been a problem. This makes it clear that when there is a man-animal conflict moment, it's not just the animal that is the problem. It all seems to come down to behaviour. If you are aware, alert, quiet, calm and non threatening, animals mostly leave you alone. Yes, of course there is always a chance of error, one cannot discount that, because it would be stupid to do so. But with my years of experience of being around various animals in various ways, I firmly believe it can be done.

Allowing walking safaris with tigers will solve two problems. One, it reduces the burden on the forest department by allowing tourism also to be a way to keep an eye on the animals. Secondly, it will get people more involved with the ecosystem, as in the walk they will learn and feel the forest and even if they don't see the tiger, a pug mark or scratch on a tree will make them aware they are in the presence of the world's largest feline predator. They will see other animals and birds instead of just rushing around in a jeep looking for the tiger. It will also teach them decorum and forest behaviour which I believe we sadly

The grace with which tigers place their paws down on the ground
as they walk is a lovely thing to see. Corbett, India.
Photo credit: Neel Soni

lack right now. In 2008, when I was in Rwanda, 20,000 visitors had
added 8 million dollars to the parks coffers. That is 32 crore rupees.
Every visitor pays 500 dollars just to walk with the gorillas. This is
excluding lodging costs, food costs and transportation. Home stays
near the park earn the local people money apart from the fact that
the guides and rangers are people chosen from local communities. In
Rwanda, they also have a ceremony called the Kwita Izina. This is the
baby gorilla naming ceremony. Celebrities from around the world are

invited to come give baby gorillas their names. A huge fete like festival with music and dance and local flavour, it is a huge crowd puller and it helps the communities come together to celebrate and they also see how appreciated and loved the gorillas are, inspiring them to protect them. The whole theme of the ceremony is to celebrate nature and empower communities.

In most of our tiger reserves, we don't even have resources with which we can bring in our local communities to tour the park. I cannot emphasise enough how many times I have been in a village a stone's throw away from a national park and have met children who have never even seen the tiger. His or her first glimpse simply cannot be one late night when they are with their goats or cows. Or when they go into the forest to collect wood or go to the toilet. That is what leads to disaster.

Which then brings us to the most crucial aspect of conservation – local community support. Today in India without involving the local communities and indigenous peoples living in and around our wild areas, we are not going to be able to protect our biodiversity. The single largest reason ironically for the disappearing wild is you and me and our lifestyles. It's easy to blame people immediately in the vicinity of national parks and forests and yes, they do play a role in the biotic pressure but they don't cause nearly as much damage as you and I do with our demands, for cars, power, homes, roads, water and so much more. If your daily living is interrupted by deer in your field, leopards in your backyard and sloth bears in your village, you will see animals very differently from how we do now. For you and me, they are beautiful, mysterious and precious. For the locals, they are an obstacle to proceeding with their daily lives. Many of the people have also been removed from their ancestral lands to accommodate the needs of a national park. The few authentic indigenous people in India today are the only people who truly still appreciate the presence of the wild on their doorsteps and support the animals' right to be in the same landscape as them.

In 2013 alone, India lost 80 tigers, 42 to poaching, and these are only the direct evidence numbers. In 2014 we lost 66 tigers and we lost 91 tigers in 2015. In 2016 we lost 51 tigers to poachers.For every seizure, many go undetected. Where are we going wrong? The answer is pretty simple – political will and personal choices you and I make in our lives.

Right now, the government of India has decided to open up previous no-go areas in rich forests of Central India to coal exploration and mining. While these mines will not directly be located in protected areas and sanctuaries, they will violate the 10 km safety area around buffer zones as designated by the Supreme Court. This was put in place to stop commercial activities in and around protected areas. Unregulated mining, many of them government-owned companies, have already destroyed vast tracks of forest lands around India and are poised to do much the same in central India. One of the largest tiger populations in India exists here and they use the connected corridors of forests in which to roam and move. Without this free and easy movement, tiger populations will become isolated within certain areas and soon, in a few generations, their genes will become inbred and not viable, inevitably signalling a death knell. Greenpeace estimates that close to 10,000 square kilometres of forests will be affected by 2030, 40 per cent of which is tiger habitat, not to mention other animals.

These forests also serve as catchment areas that feed into major rivers and provide several valuable commodities like fruits, medicines, tubers, fish and firewood to local communities who will be displaced by the mining activity.

Coal fired power plants which these mines will supply are some of the most heavily polluting in the world and are the world's single largest drivers of climate change. The mines are also open cast mines and the fly ash and dust and debris destroy the immediate environment. I have walked through areas of forest before and after mining has taken over and I cannot tell you the damage I have seen. Trees look black, there is sludge in the water and the air is so heavy with toxic fumes, it's difficult to breathe. In just one hour, my clothes were dark grey and I had a fine layer of dust over my skin.

The big question is why are we in India relying so heavily on coal for our power needs? Coal is not a renewable resource. We will run out soon and then what?

The Planning Commission, the country's official think-tank till 2014 had projected that 90 per cent of electricity would come from coal even in 2030. Recently, Coal India announced that it would expand its domestic production from 435 million tons in 2011 to 615 million tons in 2017. For this to happen, more areas will be opened up by mining, causing fatal fragmentation of already fragmented and interrupted landscapes that connect our tiger-rich reserves.

Already according to a survey done by Greenpeace India, over one million hectares of standing forests of which 70 per cent is dense forest falls within the existing 13 coal fields. This central Indian landscape encompasses Madhya Pradesh, Chhattisgarh, Jharkhand, parts of Orissa and Bihar and Maharashtra. Over 10 per cent of this land shows the presence of tigers and over 30 per cent lies within the 10 km safety zone of the buffer.

The 13 coalfields already impact over 8 tiger reserves with the loss of forest connectivity and coal mining pollution and any expansion stands to do further damage. Over 230 tigers are found in just these 8 reserves without taking into account the transitioning and dispersing animals that live outside the reserves in areas the coal mining is affecting.

Conservationists estimate that over 100 people are either attacked or killed or suffer indirect losses every year, like cattle kills to tigers and leopards. Numbers could be higher. In the last year there have been close to 1500 cattle kills in Maharashtra alone in and around the tiger parks. In 2015 there were two sensational man-eating cases with a tiger in Ooty having killed 4 people and another in Moradabad having killed over ten people who is yet to be caught. Several animals have also been poisoned, electrocuted and have fallen into open wells and hit by moving traffic. Cattle compensation or payment for cattle killed by tigers to the farmers who have lost their cows goes a long way in preventing this conflict, but payments are delayed, the process of receiving the payment is complicated. Often the farmers get no compensation and in retaliation they poison the dead cow to kill the tiger. This not only kills the tiger who killed the cow but in the case of a female tigress with cubs, it also kills her cubs. Other predators and scavengers who feed on the carcass once the tiger is done with it also die. A camera trap on such a carcass once showed wild dogs, a leopard, two sloth bears, two jackals and several mongoose. If that carcass had been poisoned, all these animals would have died or been incredibly sick.

I remember how once, as a young girl, I had been standing under the shade of the ancient banyan tree in the Theosophical Society, when uncle Siddharth asked me to hug the tree. He said to me: 'Close your eyes and just hug it. Can you hear it speak? Feel the trunk beneath your cheek and the warmth under your hands. Listen to the rustle of leaves and feel all your worries just slip away. You can talk to it you know? Just remember Swati, so many people who have stood under this tree have died, as will we, and here it still stands, providing shade, breathing oxygen into the atmosphere. And when it dies, it will still live on precious because it will one day become a diamond, or fossilized fuel, when you and I will just be the soil on which it will stand.'

That's what coal is. Dead plant matter accumulated over a time period of millions of years. When confronted by Ashish Fernandes of Greenpeace who co-authored the report on coal mine threats to the central Indian tiger landscape, the Madhya Pradesh coal minister asked him if India wanted tigers or coal. As if the two could not exist together. Perhaps as shocking as that statement is, it is the truth. Both cannot exist together. But is that a viable choice for us?

No. India has the ability, if political will can focus and plan, to explore the use of renewable energies that are non-polluting to meet our needs. Renewable forms of energy include solar and wind power, geo thermal energy and others. It could not only meet our needs but also create new jobs and stimulate the economy. Most importantly, it will allow us to depend on a source of energy that we are never going to run out of. If we continue on this path with coal, petroleum, oil and diesel, we are bound to hit a wall and hit it hard and then no magic in the world is going to help us. Populations are still growing with our insatiable needs to consume. It is you and I as we stand today who threaten the tiger even more than poachers. By destroying their forests, we will also make it easier for poachers to get to them as they will have to leave forest areas and move through inhabited areas creating conflict.

As the sun set, gilding Bandhavgad fort, I watched a female tigress, the mother of the four sub-adults I had seen earlier in the day, recline her body into the grass and heard the powerful call of a male tiger sound from across the dense forest, it seems inconceivable that we could one day watch the sun set on the last tiger in the wild.

Today, if with all of our issues from growing human population, development projects, degrading and vanishing habitats, we still have tigers and leopards and other animals in the wild in India, it is not because of good management but because of a culture of tolerance and respect for the wild. A leftover echo of a feeling that was the bedrock of a widely pantheistic culture.

And it's this respect and love of the wild that is being eroded by a new and bold world of technology and needs. It's simple enough really speaking – we exist because the wilderness does. Perhaps that is why our ancient traditions accorded the wild the respect and the love it deserves. There is a saying in Africa, it's called Ubuntu, literally, the philosophy of 'humanness'. The fundamental underpinning is 'I am, because we are' and this, I believe, includes the natural world around us. I am, because everything is.

If you know wilderness in the way that you know love, you would be unwilling to let it go... This is the story of our past and it will be the story of our future.
                                              – Terry Tempest Williams

2

LIONS
THE KING AND HIS QUEENS

I love watching the Serengeti, the way lions live. The only way the king lion loses his crown is by somebody physically defeating him.
— Ray Lewis

IN FIFTH grade, I would read a book that would change my life. I did not know it at the time but it had a profound impact on me. The book was Joy Adamson's *Born Free*. I would later see the movie *Born*

Free, where actress Virginia Mckenna played Joy Adamson. Virginia Mckenna was so inspired that she started the Born Free Foundation that helps with conservation of wild lions and the rescue of lions in distress to this day. In the movie, there is a scene where Joy and George Adamson discuss if Elsa the lioness could ever truly live in the wild as a wild lioness after being hand-raised and loved by them.

> GEORGE ADAMSON: ...she can't make it. She can't think. She can't mix with her own kind... She can't do anything the wild animals do to survive. You've done too good a job on her. You've made her tame. It's too late to try to let her go wild now. All we're doing is making her miserable, torturing her. How could you be so cruel?
>
> JOY ADAMSON: You keep quiet George.
>
> GEORGE ADAMSON: I don't know what goes on in that head of yours anymore... What's wrong with a zoo, anyway?
>
> JOY ADAMSON: Nothing. Except that she won't be free.
>
> GEORGE ADAMSON: And is freedom so important?
>
> JOY ADAMSON: Yes, yes, she was born free and she has the right to live free. Why don't we live in a more comfortable setting George? Other people do. We chose to live out here because it represents freedom for us. Because we can breathe.

And that to me said it all. 'Because we can breathe.' Of course as a ten-year-old, when I saw the movie, I loved it for what it was and then read the entire series of books on Elsa with *Living Free* and *Forever Free*, following her story as a wild lioness and then her cubs. I cried when she died in *Living Free* and cheered when her cubs went on to become wild lions in *Forever Free*. I had dreams of running off into the African bush and raising abandoned lion cubs and setting them free. It would be years before I realised that maybe one in a hundred lions could be raised like Elsa in a home and then be set free. As an adult, I also realised the futility of wanting to rush off into the African bush. But I did not ever forget that the essence of life was to be born and live free. After all, as a child born in 1972, I was a child born in free India. I did not know the world my grandparents grew up in and fought when India was a British colony. I took my freedom and my rights for granted and as Homo sapiens, we do reign in dominion over all

The Dance of Death. The lead huntress agitating the angry buffalo, hoping to get them to run in panic to isolate one from the herd. Savuti plains, Botswana.
Photo credit: Swati Thiyagarajan

other fellow beings on this planet. You may even say we have colonised them. I did eventually find my way both into the wilds of India and the African bush. My role, however, would be very different. As an adult now, especially an adult who has spent two decades travelling in the wild and observing various programmes with big cats, I see the Joy Adamson method as problematic. Lions raised so close to people will never ever adjust to being fully wild and they will be a danger to other humans who cross their path as the cats have lost their inherent fear of people that keeps both the cat and us safe. Of course there will always be the individual case of success, but it cannot be translated into a methodology for saving big cats.

One dark night, with just a full moon to light the forest, I found myself in a tent in Botswana's Savuti Park, which is part of the Chobe National Park. My tent was large enough to fit me, my husband and our bags. Half my tent was made out of mesh through which I had a clear view of the surroundings outside and the other half canvas.

At about 2 a.m., I was woken by my guide, who was standing outside his tent with just a torch – he told us that we had lions in our campsite. I came immediately awake and rolled over to look out through the mesh and I saw ten lions walking past my tent. I expected them to keep going, but eight of them broke away from the lead pair, a lion and a lioness, and made a beeline towards my tent.

Now let me tell you that the sight of eight lions, wild lions, walking towards one is not something you can ever be prepared for. I could see from the curiosity and mischief on their faces that these were sub-adult males and females, probably just over two years of age. The males had only small tufts of hair and no manes. While huge, I could see they had still a little bit of growing to do. Clearly, the two lions who had walked away from the camp were the adults and these chaps did not have either the self-preservation or the dignity yet to ignore us.

The first item that succumbed to their curious questing was the little oil lamp in front of my tent. While one of the females played with that, the others split up and streamed around my tent. Two of them stayed on the side while the others went around to the back. I could then feel lions against my feet as they had now come to lie up against the canvas part of my tent where my feet stuck out of my sleeping bag. I held absolutely still, barely breathing, because having grown up with house cats I knew that any sudden movement on my part would cause a cat reaction which would involve at the least, swatting and at the worst, pouncing.

Then I felt them get up. While still recovering from that, I watched the female knock the oil lamp over. I now wondered if the grass would catch fire and then my tent. Luckily the flames went out. She then pounced on the leather water bucket outside the tent and ran off with it. Now, I could not see any lions at all but I could see my guide still standing outside with just his torch. He was holding them off with just his voice and the light. He sounded so calm that I knew that we were not in any immediate danger. Alwyn Myburg has been a wildlife guide for over a decade and is an absolute genius in the field. He has been around hundreds of lions in hundreds of different situations and really knew their behaviour well. I took my cue from him and decided to relax. Plus, I knew that the more nervous I felt, the more of that energy the lions would pick up. If I stayed calm, they would stay calm too.

However, now that all of them were behind the canvas half of my tent, I could not see them anymore. I could hear them and occasionally feel them tug on the tent and somehow it felt more nerve wracking to not be able to see them.

Craig, my husband, unzipped the tent flap for us to be able to peek out and see them. Now I could see them dragging the canvas ground cover away and rip into it while a few others were still playfully stalking Alwyn. Every time he asked them to back off and flash the light, they would back away. One of the females suddenly leapt towards us, as she must have seen movement near the tent flap. I have never zipped up anything so fast in my life. I did get some nice photos with my phone though.

Soon, we started to feel the whole tent shake as one of the males started tugging on the ropes. I started getting nervous again because if he tugged hard enough he could bring the whole tent down on us. After all, a nearly fully grown male lion has the strength of twelve men.

Under the fallen canvass, Craig and I would be like two wriggly creatures and that would trigger the cats need to pounce and play which would quickly turn to a killing mode. Alwyn, who by this time realised that the lions were getting more frisky, had reached the car. He started it and drove towards the tent flashing the headlights. This made the lions back off but it did not chase them away.

Just as we were wondering how much longer the lions would stay, a faint sound rang through the clear night air. It was the sound of a lioness calling. In seconds in front of our astonished eyes, the playful sub-adults just vanished. They dropped everything they had been playing with and ran in the direction of the call. Clearly mum's or their aunt's call was the bigger deterrent than our car.

Sleep was futile for the rest of the night. It was awe and not fear that kept me up. I knew that what I had experienced was a once in a lifetime moment. I could come another hundred times to Botswana and camp in the wilderness and chances of having lions come this close to our tent would be nil.

These lions had not come to harm us, or attack. They were being curious and playful as are all cats. For close to two hours, I had been in the presence of one of the most magnificent and formidable predators in the world. For that time, in that space, my entire being was just engaged with the power of the wild.

Briefly, I was not me. I was not connected to my house or cars or anything from my modern human life. For that time, I was as present or as vulnerable as our ancestors who chose to sleep beneath the stars,

Dead elephant. I have never seen more dead elephants than on the Savuti plains in Botswana, home to the elephant killing lions.
Photo credit: Swati Thiyagarajan

around fires and in caves. For that time, I lived in magic and I lived more than I had in years.

This to me is the greatest gift of the wild. This ability to take you out of yourself and introduce you to who you truly are in your original design. A creature as wild as the lion, who belongs to this planet and feels its rhythms and realities in the blood. A creature connected to the web of life as naturally as the stars that shine in the night sky and the butterfly that drinks from a dew drop. For that time in finding the real you, you feel whole. You feel like the human animal should. This is what we stand to lose when we lose the wild, when we forget what it is to be truly free. We stand to lose the best part of ourselves. We have put ourselves in cages of modern constructs and forgotten that we need to 'breathe'.

Lions like tigers have lost over 80 per cent of their home range in the last 150 years. It seems the world has become too tame a place to tolerate predators anymore.

An interesting fact is that until about 10,000 years ago, the lion was the most widespread large land mammal after humans. They were found in most of Africa, across Eurasia from western Europe to India, and in the Americas from the Yukon to Peru. It's also interesting that they started to vanish not when humans lived as hunter gatherers but when we turned to agriculture. The Industrial Age sealing their fate as it has sealed the fate of most of the ecosystem on the planet. The invention of the gun wiping out hundreds of species, as man slowly found that with this one weapon he could assert authority not only over his fellow man but over every other fellow being.

Today, the lion is listed as a vulnerable species. It is found only in South and Sub-Saharan Africa, having become extinct in north Africa and everywhere else in the world, except for one small population in Gir, in Gujarat, India. The biggest reason for this decline in lion numbers is conflict with people over livestock. When lion kill cattle, people retaliate by killing the lions. With most governments not including local communities in their conservation plans and private players cutting local people out of any revenue from direct tourism, a dead lion is worth more to a village than a live one. Added to this, extensive habitat loss has made the situation worse.

Right now, there are only six areas in Africa that are large enough to sustain viable lion populations. Everywhere else people will either learn to live with them or we will lose them. Lions move over vast tracts and require large prey bases to sustain their prides. Right now, they have no future without well managed landscapes with protection.

In Africa, it is estimated that there are about 20,000 lions. This is a rough estimation or rather a guestimation, as no true head count is possible. Many lion experts say that the numbers are far less. The truth is lions have seen a major population decline in the African range of 30-50 per cent every two decades since the second half of the twentieth century.

The west African lion population is critically endangered.

In this mix, without knowing if existing populations have any hope of surviving, trophy hunting of lions is allowed in several parts of Africa. An estimated 600 lions are killed every year by trophy hunters. 60 per cent of these trophies are sent to America. Trophy hunters want to kill large robust male lions as the mane is of utmost importance to them. This causes destabilisation of existing prides and is a counter evolutionary threat. The large robust male lion with the lush mane is the top of the lion genetic line. It is crucial that he survives because his survival ensures that the best lion genes are carried forward in the future generations.

In recent years, there has been a large increase in maneless males that can be directly attributed to this selective hunting. The removal of a big male in his prime is also problematic in that he could be the head of a pride and his death causes upheaval. New male lions try to take over the pride, and kill existing cubs. Lionesses get hurt or killed in the fight to defend the cubs. If the hunted male is a sub-adult just out of the pride, and looking to take over a new one and then killed, it is equally problematic as it stops all hope for future generations.

Lions by nature are group animals. They are social and need each other to survive. The lionesses do the majority of the hunting while the males assist in the take down of really large prey like African buffalo and hippos. Sometimes, two or even more brothers stay together and rule a pride. In Botswana's Moremi National Park, there is a coalition of three males who are the 'Big Boys'. These boys are unchallenged and the kings of their domain. They specialise in hunting hippos.

Hippopotamuses can get up to weights of 1,500 kilos on an average and attack and kill more people in Africa than all the other animals put together. These three boys of the Moremi catch them when they are outside of the water and kill them by running at the hippo at full speed and slamming into its neck to break it. This has to be timed perfectly to avoid the slashing hippo teeth that can grow up to 1.6 feet in length. The hippo is also pretty fast on land and can outrun most humans. When we saw the 'Boys', they had just come from a kill, their manes and faces almost black with blood.

Nature has designed male lions to rule in their prime for as long as they can if they are strong, and then allow for a younger male in his prime or a stronger male to take over. Most male lions leave the pride they are born in to take over other prides. In this way, nature keeps genetic inbreeding to a minimum. An incoming male is a threat to existing cubs and he will kill them to bring the females into heat and spread his genes. This ensures that new healthy and strong genes are carried forward. It seems like a brutal system and, yes, it is by human standards, but it is their way and it works for them and it also keeps an inbuilt check on numbers. Lions as apex predators are far smarter than us in their realisation that prey must always outnumber predator. They figured that they must never outnumber their food source. A simple lesson that we humans have not been able to master.

Balance is key to the survival of all ecosystems and except for homo sapiens, every other animal that is wild and not either domesticated or a pet adheres to this fine balance. Trophy hunting and illegal killing of the lions throws out this fine balance.

Trophy hunting is still justified by governments who insist that it brings millions of dollars in revenue to the local people. There has been no independent audit to show any proof of this. Even pro-hunting organisations like the International Council for Game and Wildlife Conservation have reported that only 3 per cent of revenue from trophy hunting ever makes it to the communities affected by hunting. The rest goes to national governments or foreign-based outfitters.

Tourism, however, brings billions into the African economy. Most of the tourists come to see the wildlife and if lions disappear from the landscape, the impact on tourism will be significant and could cripple the economies of several African countries.

'Maybe one day I will grow up to be like my warrior mamma?'
Lion cub with mother, Savuti plains, Botswana.
Photo credit: Swati Thiyagarajan

There have been attempts to introduce sustainable methods of trophy hunting but lion populations continue to decline. One of the better ideas was to only target ageing male lions who have already contributed to the genetic pool and are now ousted from a pride. The verification of this method fails though, as it is only done after the animal has already been shot.

The biggest problem is that the African lions are listed as vulnerable and not as endangered, which could bring them under the protection of the Endangered Species Act and stop active trophy hunting, but this is being fought by governments and wealthy private hunting outfits.

One of the most insidious practices that started up as a 'solution' to trophy hunting is the canned hunting of lions, a phenomenon that I became familiar with only after I came to South Africa.

On a grainy screen, I watched as a lioness was snarling, backed up against a fence. On the other side of the fence were her three cubs. Suddenly some people come into view in a jeep and they just shoot her. Bullet after bullet strikes her and she convulses and dies. Another clip shows me a lioness in a tree and below her is a jeep. People in the jeep hold cross bows and then just shoot the bolts into her. It was horrific, macabre, and it undid me. Canned lion hunting is the practice of shooting and killing lions and other big cats that have been captive bred by breeders and when they are old enough, they are targeted by hunters. The reason it is called canned hunting is because the animals are not in a reasonably sized area where they can feed themselves, roam free and live naturally, where they have the sense to flee and hide. It refers to hunting that is done with fences. Either a physical fence that stops the animal from being able to escape or mental fences, wherein due to the fact that the animal has been raised by people, he does not see people as a threat.

I was doing a series called *Born Wild* out of Africa for NDTV, and was doing research on lions in Africa. In the process, I came across this phenomenon of canned lion hunting. I immediately called International Fund for Animal Welfare (IFAW) and World Wildlife Fund (WWF) to ask if the practice actually existed and how widespread it was. To my horror I learnt that it was very real and very widespread. I immediately decided to do a two-part episode on lions because I felt that we really need to know more about this practice.

In the course of my investigations I found Simon Tricky. Simon Tricky, a guide and ranger, was the man who blew the whistle on canned lion hunting in South Africa. It was after Simon's undercover footage was widely broadcast on TV about 14 years ago that South Africans in general even heard about the practice. For his troubles he has been threatened, shot at and attacked. Simon gave me a really clear look into the practice and it is disgusting. He calls it the 'Rack 'em, pack 'em and stack 'em' practice. Lions are bred in captivity. Cubs are removed from their mother almost as soon as they are born so as to bring the females into heat again. Female lions are sold at really low prices to other breeders or hunters without big pockets who just want the thrill of hunting. Males are the big prize. Once a male reaches an age where he develops a mane, he is immediately targeted for hunting. The targeted animal is then drugged and laid out in an enclosed area. As he has been hand-raised in captivity, he has no natural fear of people. In fact, he sees people and associates us with food. A lion raised in captivity will not run when he sees a human and certainly

One of the naughty sub-adults who visited my tent in Savuti.
That was my ground sheet.
Photo credit: Alwyn Myburg

does not recognise a gun or a bow and arrow. In any case, he is drugged and the hunters are brought to him by so called guides who pretend they have tracked him down, and then as he comes to and is groggy, he is shot.

Advocates of this practice claim that by shooting captive bred lions they are decreasing the pressure on wild lions for their trophy. They also claim that by breeding these animals in captivity they are helping the ecosystem by raising animals that can one day be released into the wild if they become extinct in places. The first claim is insidious. The first claim could work if the only hunting that was practiced was canned lion hunting – morally reprehensible in practice but not reliant on wild populations. But as these outfits are privately owned the government has nothing to do with them and, consequently, cannot decree that only canned hunting be allowed. The second argument is patently false as these captive-bred, human-raised animals can never be re-wilded. One, they have no fear of humans making them dangerous if released and two, they have never learnt to hunt for themselves, so either they will die unable to source food or they will kill cattle or a human and be shot for it.

A public outcry against this practice then prompted a government enquiry into the matter which became another joke. It turned out that the government was already aware as none of the outfits can operate without a license. In many cases, licenses were procured purely through bribes. The animals are often housed in deplorable conditions, especially the females, who have no value except as breeding animals. The upkeep of predators is a costly business and their owners want a quick turnover on their money, so the breeding is ruthless and they are farmed like factory chicken.

Instead of banning the practice outright, the government then just brought in rules of conduct they said would help regulate the situation. The main rule being that the animals had to be housed in enclosures that would be large enough for them to live and hunt in and that they must be self-sustaining. An eyewash, as there are too many hunting outfits to monitor and well placed bribes helped turn blind eyes. Even this was fought in court by several hunting outfits, just one of which had a 90 million rand turnover from canned hunting. With that kind of money involved, it became clear that this practice would not disappear in a hurry. Only an outright ban of all breeding and housing of predators and hunting could possibly make a dent in this practice.

In itself, captive breeding did not start out as an evil. It started out with the noble attempt by zoological societies to breed animals that were endangered in the thought that at least these populations would be safe if there was worldwide extinction of the species. Many species today actually exist only in zoos while they have died out in the wild like the New Guinea singing dog and the Pinta Island tortoise. The fact is that certain species can be captive bred and released into the wild. These are species that are naturally not dependent on parental care to learn how to survive. These are species that usually do not form attachments to their human care givers as they have never been nurtured like mammals are by mothers and fathers. Even non-mammalian species like birds who get close parental care struggle to survive once raised in captivity. Some species of birds do manage to adapt and survive but reptiles, butterflies and insects are probably the best at it.

Lions, like our tigers in India, have a white counterpart. The white lions are found naturally only in one area of South Africa –

the Timbavati. For some reason, lion populations here have always thrown up a gene that leads to white lions. These animals, like our white tigers, are not albinos but the result of a recessive allele. They are very rare. Two tawny lion parents can have a white lion cub, while two white lions will only have a white lion cub. Because they are so rare in the wild, they are a favourite of breeders. Zoos want them, private owners want them and certainly, trophy hunters want them. Breeding with white lions is a terrible idea because within a generation or two, physical problems from blindness, weakness of limbs to hip displacia and a host of other problems take over. The recessive allele that gives these animals their white colour is not a healthy gene and leads to very quick and apparent inbreeding issues.

I was lucky enough to get to know two white lions and learn about all of this. Tsau and Tendile, a white lion and lioness, were rescued by my friends at the Jukani Big Cat Sanctuary, when they were just cubs, from an unscrupulous breeder. Tsau would have been sent in for a trophy hunt and Tendile would have been kept to breed more cubs. In nature, the white lion is a rarity as they are a result of a recessive gene. Their white colour does not blend with the landscape like the tawny colour and it makes it harder for these animals to be stealthy and effective hunters. They are sacred in African indigenous cultures and are believed to be the children of the stars. Legend says white lions are the guardians of the human soul and the more evil there grows in this world, the fewer white lions are born.

I met Tsau and Tendile when they were about six months old. Already quite big and strong, they were absolutely enchanting. Having been hand raised and brought up in my friend's house with his cheetahs, the lions were friendly and curious, enjoying being around humans. They were kept away from the regular visitors for a long time and only moved into their own enclosure when they became adults. At Jukani, they get life time care, will never be used for breeding and are ambassadors for their fellow white lions who are mistreated in captive facilities around South Africa every day.

Thanks to Tsau and Tendile's great patience, I learnt to be around large cats and got an up close and personal look at those jaws, paws and claws. It also firmly established for me that animals are individuals and I can never look at them in a remote scientific way that views them

*Another sub-adult from the group that visited me in the middle of the night,
contemplating my lamp, outside my tent.
When they are young lions are incredibly curious and playful.*
Photo credit: Alwyn Myburg

as 'species'. I have played with them, napped with them, fed them,
walked with them and kissed their faces. It will never ever be anything
but personal to me, the fate of lions on the planet. To me it will always
be a deeply horrifying loss like the loss of a family or friends. Indeed,
it might be something more, because there is a precious divinity to the
trust that wild things bestow upon you and imagine violating that trust
with a bullet.

Today Tsau and Tendile are six years old. Adult lions in their prime
and I see them rarely and while I can't romp with them as I did years
ago, I do get the occasional hug.

Which was why it was even more horrifying to make one other
connection. In Africa, in private game reserves and zoos and sanctuaries,
people are allowed to go in and pet the animals. Tourists flock to these
places because, of course, everyone wants a picture with a lion or a
tiger or a cheetah cub. The question no one asks is how is it that these
places always have cubs? Where do they come from? Where are their
mothers? What happens to them once they get too old for petting? I
asked these questions and the answers are shocking.

Two old boys on their way out as the reigning kings, feasting on a lucky kill of an already injured springbok. New males were already challenging them for territory.
Kwai Conservancy, Botswana.
Photo credit: Craig Foster

The cubs come from breeding farms. They are removed from their mother when they are very young so as to make them used to people and allow for humans to come in and pet them. With animals like lions, the window period for petting is probably four, maximum five months. Once they become too big to be safe anymore as even a captive bred animal can and will be ferocious, they are sent back to the breeding farms where they are used in the breeding cycle and worse, the males are sent off to be hunted. So a tourist thinks he is having a lovely moment with a baby animal, whereas he is only really contributing to its miserable life and ultimate death.

There is a whole involved argument on human-animal interaction and why we feel compelled to want it, and why, sometimes, the animals even enjoy it. I cannot lie and say that my personal contact with animals was not precious. It changed deeply inside me the way I viewed them after, even in the wild. Having lost our own wildness, we humans feel an innate need to connect which explains our obsession with our pets. I totally understand why children and adults enjoy that contact with an animal. I know that many animals too enjoy this contact but mosty it is coercion. A cub does not get to have a choice. He or she is just stuck with hundreds of people coming in and petting them. It can get very stressful. If they snap and attack, they get put down. More and more we know that animals in captivity are dealing with a lot of stress when not treated right. I will never and have never gone to a commercial petting place. I have been lucky enough with my work to have had contact with these beautiful beings.

For the most part though, it's a shocking practice and should be banned. For one, what I have already said above about the reality of what the presence of cubs for petting means. Two, there is a lot of stress the animals take on with all this petting. Imagine your own pet dog. Would you want hundreds of strangers coming into his enclosure and petting him? So why is it ok to pet a captive cub?

There are some places that do not breed their animals and occasionally, when they acquire cubs, they allow informed petting and only do so because all these animals will get lifetime care with them. Which means the cubs will be raised and looked after the rest of their lives in that place. It is still not ideal but better than unregulated petting.

There is another great danger to wild lions in Africa. In the wild in Africa, there are very few lion populations today that are free of disease. Tuberculosis is the biggest problem. The ungulates they kill carry the TB germs and the lions pick it up from the meat. With many of the lions also carrying the FIV or the feline immunodeficiency virus, TB is making lions across Africa much weaker and reducing the reproductive rates.

At numbers like 20,000, it might seem like this number is almost ten times that of the tiger numbers world over and therefore marks a healthy population, but these numbers are distributed across a continent and very few countries have more than 2,000 wild lions. If the tiger is the symbol of wild India, then one is not overstating that even though Africa is known for its Big Five, it is the lion that stands as a symbol of wild Africa. No, they are not in the same critically precarious position as tigers are right now, but they are facing challenges that if left unattended can bring them to this critical position in a decade.

There is one population of the lion that is critically endangered, and that is the population of the Asiatic lion in Gir, in Gujarat. There might be a debate today in the country about whether lions are indigenous to India at all, but the reality is that they are here and they need more protection than they have right now.

It was a hot three days. The middle of summer, the heat behaving like a real character all by itself, clubbing unwary travellers over the head as they made their way towards the Gir forests in Saurashtra. It gleefully kicked, bit and stomped on you. Feeling rather flattened and battle scarred, Gargi and I had just spent a day wandering the forests, looking for lions. We were here to do an episode on the lions of Gir, one of our firsts for *Born Wild*. Admittedly, summer was not the best time to do so.

We just had another day and a half left and were, of course, very aware of the fact that we could hardly do a half-hour episode on lions without lions. The following morning, we were at the gates by 5 a.m. With temperatures reaching an astonishing 50 degrees celsius and the glare becoming too difficult to film in by 10 a.m., we had small windows of time in which to film. The forest department had very kindly asked for several forest guards armed with walkie talkies to be on the lookout for lions and to inform us as soon as one was spotted. As hour after hour melted into each other, we grew quite stressed; we had just that one day and the next morning at our disposal to complete our shoot. Our budgets were small and we could only afford to stay for a short period of time to complete our filming. It was always challenging to be able to shoot and compile enough material for our half hours in the time we had. We were also very aware that we had been given a budget to do this programme, a budget that might have been better utilised by regular news and that NDTV was encouraging us only because the channel had a true concern and passion for wildlife. So it was imperative that we managed a programme and did not just waste that money. We were also well aware that it was only NDTV that also encouraged us to do a long format wildlife programme for the news.

It was going to be *Born Wild's* first run on TV and both Gargi and I wanted it to be a success because if we could pull it off, we could then do more seasons. In the time we had not spotted the lion we had focussed on all other components in the story.

The Gir Asiatic Lion Forest Reserve was originally established in 1965. The forest area of Gir and its lions were declared as 'protected' in the early 1900s by the Nawab of the princely state of Junagadh. At the time, there were less than 20 lions left, the rest having been killed mostly by trophy hunting. Today, the Gir sanctuary covers a total area of 542 sq miles of which about 100 sq miles has been made into the national park and is home to over 400 lions. Recently, a group of lions wandered out of the Gir forests and took a long walk in Junagadh city both alarming and exciting residents who could not believe what they were seeing. Over the years residents here have seen the ocassional lion on the road, but this time as the sun set and darkness fell over the city, over 8 lions took a walk through the lanes. No cattle were killed and no human was attacked. The group then made their way back to the forest. Maybe they were like our 8 lions in Savuti, curious and playful but not threatening. However it is a situation fraught with the potential for some serious conflict.

And it was into this space that Gargi and I had arrived in the hope of seeing lions. On our last day there, we were hot, weary and, now, close to panic. Forest guards had no word on the lion and the sun was climbing in the sky. Shortly it would just become too hot and too bright to shoot. We stopped at a water hole to get a few shots and I stayed back in the jeep to get some water for ourselves.

There are seven major perennial rivers that run through the Gir region. Saurashtra is a dry region without major rainfall even in the monsoon season. Gir is the main catchment area here for water and a lifeline for the people. At one time, population numbers in Saurashtra were low and farming was not a major occupation. All crops grown were low water intensive crops. With sugarcane becoming a major cash crop, Saurashtra too joined in the sugarcane boom. Sugarcane is a crop that requires a lot of water and soon, artificial means had to be put in place to get enough water to the farmers, from the controversial Sardar Sarover dam on the Narmada and four reservoirs on four of the Gir rivers.

The waters are all pumped out of Gir by a growing human population, leaving the park more and more parched. Now over 300 water points dot the park in order to bring water to the animals in the summer months. Water holes have always been a controversial topic in conservation. On the one hand, it is hard to debate that summers are harsh in most Indian forests and water holes do provide the animals with water. However, nature intended for the forests to function without human interference and it is because of human need draining the forest of its resources that water needs within the forests for the animals have also changed.

Funds are given to most parks every year for conservation purposes and often, these funds are misused in the building of unnecessary water holes. I have seen how some water holes have been of great help and how some are nothing but a scam. At the end of every year, the forest department of every park has to account for the funds and spend them in order to get the next year's allocation. Water holes are an easy way to spend part of the money and misappropriate the other half. The money would be better used in paying salaries, hiring more guards, and investing in better equipment for protection, but these are not the areas where it is spent. The department usually claims that the money went into water holes and eco development projects for the local people. Certainly, 300 water holes seemed a bit excessive to me.

Gargi and our camera person were at the water hole. Our jeep was parked off to the side of the road atop a road that sloped downwards. From here I could see the whole of the valley in front of me. To my right were gentle sloping hills and to my left, dense forest. The whole of the forest was a hazy gold. Gir is the largest dry deciduous forest in western India. The dry scrub, grass and teak jungles are occasionally broken up with green riverine patches. These riverine patches surround perennial streams that run through the forest and feed the rivers. Gir is a cornucopia for the region, providing nearly 5 million kilos of green grass by annual harvesting which is valued at approximately 50 crore rupees. The forest provides nearly 123,000 metric tonnes of fuel wood annually. The grass harvesting and fuel wood collection has its own adverse impacts on the environment. The ungulates, that is the deer population, depend heavily on this green grass for their feeding, and the collection of wood and harvesting activities causes huge disturbances with human activity in the forests. A growing population has also ensured that this natural bounty offered by the forest is no longer sustainable.

As I stood near the jeep and contemplated all of this I had a mental flash back to the old days. To a time when lion counting in Gir was done physically. In that it was a count by sight of the population.

Watching their mother challenge wild buffalo.
It is a dangerous learning curve. Buffalo herds are aggressive and
will fight to the death to save one of their own. Savuti.
Photo credit: Swati Thiyagarajan

Tigers will slink through the forests unseen and hidden while lions stand their ground confident and in plain sight. The reason we had not seen any was not because they were hiding but because the heat was driving them into the cool recesses of the inner forests heavy with trees and shade. In the old days, live bait in the form of goats were brought in and the lions were counted while they emerged to feed. It has been decades since this form of counting has been practised because it was seen as something that habituated the lions to people and associated us with food, never a good idea with a carnivore. An older forest officer who, as a young man, had participated in the census, told us a funny story. He and a forest guard were taking a young goat to tether him to a tree near a water hole. While they were walking the goat between them, a large male lion came up behind them silently, hooked the goat with his paw and took it away while they were still standing there. It took them a few minutes to comprehend what had happened and get their legs to work. When they did recover, it was to get away as fast as they could from the area.

The Asiatic lion definitely is a much gentler cousin of its African counterpart. That does not and will not preclude it from attacking a human, but the attack will happen only after a lot of provocation. In decades past, very few people have been attacked or killed by lions. Livestock, on the other hand, have become fair game.

I decided to take the water to Gargi and our camera person. As I slid off the jeep, I saw some movement from the corner of my eye. I saw a huge group of langurs running across the road, crossing from my right to my left and disappearing in the jungle. Then I saw another troop do the same. I turned and looked towards my right and on the small hilly slopes, I saw a lion streaking through the trees. Behind him came some forest guards. Obviously they had flushed him out of wherever he had been sitting. In a split second I saw what Gargi and Gautam could not. I saw the lion heading down the slope directly towards them. I dropped the water bottles and started to run down the road, calling Gargi's name, trying to warn them about the lion. Gargi suddenly looked up and saw the lion and indicated to the camera person to take shots. The lion who was almost upon them suddenly realised the presence of these people and he veered in another direction, heading towards the scrub jungle.

Gautam, our camera person, on seeing the lion heading directly towards him, had done the sensible thing and stepped away towards the forest guards who were with us. Gargi picked up the camera and pushed the tripod towards me. The two of us took off in the direction the lion had taken, intent on getting some shots. Sharp thorns pulled at our skin and caught in our hair. We had to push through dense scrub and low lying trees to keep the lion in sight. A few minutes later, Gargi came to a dead halt and I crashed into her. 'Swati', she said, 'why are we chasing a lion?' That's when it hit me, the huge risk we had taken! In the heat of the moment, with the worry of the episode looming large over us, we had not thought of anything except getting shots of the lion. Here we were, standing in the middle of a dense jungle, not able to see where the lion was and putting ourselves in the exact position one must never be in with a wild animal, which is to be unable to see it and making it nervous. We backed out slowly and literally did not breathe until we were back near the water hole.

In the fourteen years since that incident, I have never ever done anything like it again and never will. There are five rules that are sacrosanct in the jungle. One, be quiet, silent and calm. Only observe and move through a wild space with great respect for what it holds. Two, always keep your eyes open, be alert, watch everything and, if on foot ensure you will always have a line of sight on the animal you are tracking and for other animals that you can't see but are around you. Three, never ever think you are in control, for it is in that exact moment you will make a mistake, and mistakes are fatal. Four, just because an animal has a reputation for not being ferocious does not mean it won't be under the right circumstances. Five, finally, always respect the animal in its habitat. Know that you are the outsider there so unless you take the time and make the effort to become as comfortable in that habitat as the animal, never ever think you know what you are doing. Patience, respect, trust, calm and love are the only way to approach the wild. I could almost hear uncle Siddharth in my head with those words.

The Asiatic lion is smaller than the African lion. They are paler in colouring and move around in smaller pride numbers as the size

Asiatic lions walking away in Gir. A mother with her cubs.
Every cub in Gir is precious but there is the danger of inbreeding.
Photo credit: Dr Anish Andheria

of prey they kill in India are also smaller than those in Africa. Their manes are shorter and they have a long fold of skin on their bellies. They are also definitely much calmer and less ferocious.

But somehow, the glimpse of that lion seemed to break the jinx and for the rest of the day, we saw many lions and acquired the necessary number of shots to complete our episode.

One of the biggest concerns over this population in Gir is the fact that they are highly inbred. In fact, some genetic analysis done on individuals has shown that the lions are identical twins, or carbon copies of each other. This shows intensive inbreeding. The problems are multifold. If there is some serious genetic disorder, then it will be carried forward with nothing to stop it. The other concern is that with all of the lions concentrated in one area, disease or natural disaster could wipe out the entire population. In 1994, in Africa, in the Serengeti, 1,000 lions were wiped out in an outbreak of a virulent strain of canine distemper, a virus that usually affects domestic dogs and other canids. Several lions die every year from TB outbreaks and feline distemper. A fatal outbreak, like the one in the Serengeti, will wipe out the Indian population of lions.

Every ecosystem also has a certain carrying capacity. This is the ability of the ecosystem to sustain life. In Gir, in the last few decades, while the size of the protected area has not increased, lion populations have tripled. In the last two years, 42 lions and 37 cubs have died. Some of the deaths were natural but the high cub mortality indicates something else. Too many adult male lions in small areas leads to many fights in order to take over prides. The death of a male lion leads to the death of his cubs. A Right To Information appeal revealed that between 2007 to 2012, over 200 lions died in Gir.

Since the 2010 census, it is now believed that there could be close to 500 lions in Gujarat.

This has led to other developments. Lions are natural nomads, which is why at one time, they were as spread out across the world as humans. The growing lion population in Gir has forced dozens of lions to move out of the park and spread out across Saurashtra. In 2006, a lioness and her cubs were spotted in Diu, an island off the Gujarat coast, 95 km from Gir. Today, the Gir lions are spread out over a 10,000 sq km area. None of the places where they have spread to are national parks, which means there is human interference, grazing cattle, roads, railways and industry.

In 2014, in February, four lions died after being knocked down by trains both in the sanctuary area and 40 km from the sanctuary. In 2015, 8 lions died in floods in the park.

Since 1995, there has been a movement by conservationists and field biologists from the Wildlife Institute of India to move some of the lions to another protected space in order to split populations and give them time to recover some genetic diversity. Many areas were assessed in and around Gujarat and rejected for their unsuitability. The one space that seemed like a viable option was the Kuno Palpo sanctuary in Madhya Pradesh, a good 1,000 kms from Gir. While the Madhya Pradesh government is on board, the Gujarat government adamantly refused to allow the transfer. In 2013, the Supreme Court ordered the Gujarat government to comply with the transfer. A six-month deadline was given for the lion transfer and so far, 3 years later, nothing has happened. The Gujarat government has appealed the order and the MP government has accused them of failing to follow through. In

The dance of death continuing with two more lionesses now joining the challenge.
Photo credit: Swati Thiyagarajan

the political wrangling that has carried on for ten years, that involve arguments like claiming that the lions are the pride of Gujarat, it is hard to say if the non transfer has been a good or a bad thing. Advocates of the transfer point out that Gir is too small now for the growing lion numbers. They point out that inbreeding is causing mortality, with most of the adult lions dying from a brain haemorrhage which cannot be explained. Many lions in Gir are also killed when they fall into open wells that surround the sanctuary area. They say unless the population is split up, any disaster, right from disease to natural phenomenon, can wipe out the population. The arguments against such a transfer are equally compelling. Prior to this idea of moving lions to Kuno, two lion tranfers were attempted in India.

The first attempted transfer was with African lions. In 1904, the Maharaja of Gwalior had cubs flown in from Africa for this project.

The cubs were placed in an enclosure and raised to adulthood. Two of the females gave birth to five cubs and the Maharaja, in great enthusiasm, introduced four pairs of lions in the state jungles of Sheopur which are near the present demarcated release site Kuno. The

area covered about 1,000 square kilometres. but it proved disastrous. The lions started to attack and kill livestock and finally, resorted to killing people. From 1910-1912, nine people were killed. The lions were captured and re-introduced in a scheduled forest area in 1915. By 1920, the lions had dispersed and created panic as man-eaters. They had to be tracked down and shot. The last of the lions, a lioness, was finally shot and killed in 1928. It can be argued here that this release was done with no research, no field biologists and with the African lion whose reputation is more fierce than the Asiatic lion.

Yet another re-introduction was attempted in 1956. This time it was with Asiatic lions and they were moved to the Chakia forests in Uttar Pradesh. In 1956, one lion and two lionesses were captured from Gir and placed in the Sakerbaugh zoo, the only zoo certified to breed a pure strain of Asiatic lions in India. The zoo, located in Junagadh, housed the lions for nine months before they were released into their new sanctuary, the Chandra Prabha sanctuary, in 1957.

An enclosure was created with a three-metre-high (9.8 ft) barbed wire fence within the sanctuary in which the lions were temporarily housed before being released in the sanctuary. Initially, the lions prospered, increasing in number to four in 1958, five in 1960, seven in 1962 and eleven in 1965, after which the population died out inexplicably.

In a study done in 1999 by Dr John Singh of the Wildlife Institute of India and Dr Chellum, one of the architects of the new move, lack of adequate space, grazing, the long period of captivity in Sakerbaugh zoo, absence of education of the local villagers and lack of conflict resolution mechanisms are listed as contributory factors for the relocation disaster.

In Gir, while there are people living in and around the forests, these are people traditionally used to having lions in their midst. The main community living in the Gir forests are the Maldharis or nomadic herders of cattle. They live in small villages called 'ness'es and though they lose cattle to lions, they have never been in conflict with them. Their presence in the park might add to problems like over grazing, firewood felling and so on, but they have never killed or been killed by the lions. I spoke to an elder in a ness and he told me that they see the lions as part of the fabric of their lives in the forests. 'They are like children,' he tells me. 'They can be good and bad. You can get angry with them but you also love them.'

It's this calm tolerance for a large predator in Gujarat that has also allowed for the lion population to grow and spread out beyond Gir. A traditional tolerance that is not there in the Kuno area. Kuno Sanctuary actually has a pretty bad history with large predators. At one time, the sanctuary was home to over twenty tigers. Over the years, all the tigers, with the exception of one or two, have vanished from the area. Kuno had villages within and around the sanctuary. While many of the villages within have been moved out, the biotic pressure from surrounding villages is huge. Many of the people living here also own guns and it has been decades since they have had to contend with a large predator except for the occasional leopard or tiger. The presence of tigers is also worrying as tigers and lions have never traditionally occupied the same spaces. These are all arguments that the Gujarat government has used to delay the transfer to Kuno. Right now there is also camera trap proof that two tigers have moved to the Kuno forests from Ranthambhor, which ironically is really good news as Ranthambhor over full with tigers really needs spaces into which these animals can disperse and since this was natural dispersion without forced relocation, the tigers have settled into the new area without a problem. It is however very bad news for the lion relocation plan. Both these mega predators have never shared ground in peace and a new relocation programme with lions cannot start with tigers present in the same areas. They will fight to the death for territory which the lion will lose.

The proposed relocation plan which has been in the pipeline since 1995 was meant to happen in three stages over a twenty-year period.

Stage one was meant to last from 1995 to 2000. Twenty-four villages from within Kuno were meant to be shifted out and the habitat improved.

The period 2000 to 2005 would involve the fencing of the lion introduction site, the actual transfer of lions, and the research and monitoring.

The third and final phase would last from 2005 to 2015 and would focus on improving conditions for the local people.

In 2005, the Gujarat government stalled the transfer of nineteen lions. The transfer of villages was also controversial as many of the villagers alleged that their compensation had been inadequate and the government had not kept its promises. Questions were also raised on how ethical it is to remove people to move in animals. This is one major reason for the anger and intolerance growing in India today towards our mega fauna, because people feel that the animals are being given more importance and being saved for the rich tourists and they themselves are being marginalised. A very valid argument and feeling in my experiences across the country. In places people can barely feed their children and have had their traditional access to forests abruptly cut off without sensitive and generous compensation packages to make their lives better, dead tigers and lions are more important than living ones. Dead tigers and lions mean access to forests that don't become national parks and sanctuaries. There has been a huge lack of education and involvement of local people within India's forest areas in conservation leading to this resentment. Today, the rehabilitation packages offered by the national tiger conservation authority to relocated villagers are in the vicinity of ten lakh rupees per adult person in a family or partial cash and land. On paper, this looks like a generous offer to people who will not otherwise see that much money in their lifetimes. The problem

Two male lions facing off with two buffalo.
It takes great skill to kill buffalo as they are formidable foes
who will congregate to help a comrade in danger instead of running away.
Photo credit: Swati Thiyagarajan

is once the villager receives the cash in a lump sum and moves, there is no one to help them adjust to their new lives or even manage their money. Their lack of experience with money or farming leaves them wide open to exploitation and corruption. Just like years of scientific monitoring go into successful rehabilitation of animals, so must there be a system that enables the villager's proper transition into a new life.

Better yet, access to education and awareness should help keep population numbers down and making them partners in conservation should enable them to understand the link between ecosystems and animals and their own survival better. An irate man once told me: 'Forests belong to us, we have lived here for years. The forest department should control its animals.' It astounded me that they saw the forests as theirs but the animals as a separate entity. That is when I realised that many of them had never seen a tiger or an elephant or a leopard or a bear unless the animal was in their village, threatening them. They also see these animals as the reason they are harassed and made to move and not allowed into the forests. There was absolutely no understanding of the fact that the animals and the forest were intertwined and would die out without each other. Their grandfathers and grandmothers understood this truth but now the rift between conservation efforts

and the people has created this enormous disconnect with nature that is destroying the very fabric of the lives we live.

The plan in Kuno is to introduce one or two free ranging lion prides and monitor them until they settle in. Several conservationists have expressed concern over how the prides are going to be chosen. Lions have extremely close pride bonds and splitting up pride members can lead to great stress for the animals. It is assumed the Wildlife Institute of India, a very well respected and world famous institution, will only move existing prides, but concerns are valid. In Africa, lion re-introduction takes place all the time. Many have been immensely successful while several have failed.

The one question that is not being adequately answered is why the natural dispersion of lions from Gir over 10,000 square kilometres has not been seen as an important new development in lion conservation. Some surveys have shown that the dispersal could be over an area as wide as 20,000 square kilometres.

Natural dispersion is always a far better way to ensure animals, especially large carnivores, establish new homes. If the Gujarat

My tent in Savuti. Just some canvas between me and the wild.
My five star holidays are the carpet of stars in the sky.
Photo credit: Craig Foster

government can create protected spaces for these smaller satellite populations while maintaining connectivity with Gir forests, it is a way to alleviate the greatest concern in Asiatic lion conservation – that all the lions would be trapped in one area and vulnerable to any disaster. Right now, all the lions outside of the Gir protected area are vulnerable to threats of accidents, poaching and conflict with people. As a state, Gujarat is very proud of its lions, a fact that has kept them alive for this long. However, such feelings of pride can also be overweening in their intensity, preventing the people from seeing what might be the best for the lions themselves.

The six-month deadline handed down by the Supreme Court is long past and even with Mr Narendra Modi as Prime Minister now, with the BJP governing Madhya Pradesh the question is: will word of law or individual pride now prevail in the future of lions in India. For now it seems the relocation idea is at a standstill. The MP government is infact proposing to go ahead with Kuno as a tiger park. It is prime forest habitat and it is too much of a waste for it to just lie there unused while governments wrangle over what to do. With the two resident tigers there now, it seems that proceeding with Kuno as a tiger habitat seems best.

The year I was in Gir, 2004, was the year the physical transfer of lions was meant to happen. As evening fell into a hushed silence while the heat of the day was still rising from the parched earth, we walked quietly behind a lioness. She soon settled down in a clearing and Gargi and I did our pieces on camera, sitting on a log with the lioness behind us. I remember saying that I wished for the queen of the jungle to reign for ever here. She was calm and her intelligent golden eyes watched us. I wondered how many cubs she had had over the years and what she had seen. I wondered if she knew she was at the heart of a pitched political battle between two states, between conservationists and biologists and between people. I only knew one thing for sure, she was a master predator, her ancestors walking the planet, just as the first hominids or precursors to our human species walked the earth. They have been entwined with human history and human presence for a million years. We call ourselves 'lion-hearted' when we are brave, we get the 'lion's share' when we take the best of something, we describe our best and brightest as 'majestic lions'. They have always represented

the best part of us reminding us of our own wild natural heritage, no
matter how far and how hard we have separated ourselves from it.

> *Something will have gone out of us as a people if we ever let the*
> *remaining wilderness be destroyed.... We simply need that wild*
> *country available to us, even if we never do more than drive to its*
> *edge and look in.*
>
> – John Muir

3

LEOPARDS

SPIRIT OF THE WILD

'I hate to tell you this,' Jason said, 'but I think your leopard just ate a goddess.'
— Rick Riordan, *The Lost Hero*

IN THE summer of 2002, while doing research on leopards for my show *Born Wild*, I met an interesting couple in Bangalore. Gajendra

Singh and Visalakshi Devi owned and operated a resort in Bandipur
National Park and were avid nature lovers. They had an extraordinary
story to tell me about leopards. They had with the permission of the
then chief wildlife warden and field director of Bandipur National
Park, hand reared two leopard cubs, Baby and Bully. What I would
learn and then see a year later changed my life.

One of our greatest primal fears is of snapping jaws and slashing
claws in the dark of the night. The fact that people, cars and the
mosquito kill more people per year than all the wild animals put
together is something we seem to be able to live with. A single kill by a
big cat, however, is enough to drive us into a state of shock.

Most of my leopard sightings in the wild have been fleeting
glimpses of a cat slinking through the undergrowth, trying to stay out
of sight. My best leopard sighting was in the buffer area of the Panna
Tiger Reserve. It was a hot afternoon in the summer of 2004 and Gargi
and I were there to actually look for tigers. As our jeep took a curve,
I happened to look down into a shallow ravine and saw a leopard
drinking water. I asked the driver to stop the jeep immediately and
my camera person, Mandakini, started to shoot. The late afternoon
sun gilded its back in a thousand shades of gold and those bold black
rossettes stood out in stark relief. I remember an old South American
shaman telling me that the jaguar was worshipped as sacred by the
Incas as they saw in those dark spots the map to the heavens. I could
totally understand that because it felt like I was looking at black stars
in a night of gold. Then the leopard slowly raised his eyes to me and I
fell into that magnetic gaze. Mandakini then signalled that there was a
second leopard in the bushes next to this one and I realised that these
were two sub-adult siblings who had yet to part company. One of them
was shy and the other bold. We held very still until they both presented
themselves to us.

They moved away slowly further down the path of the shallow
ravine and I could see that there was a small path that would take us
to the edge where we could film them better. Taking permission to get
off the jeep, we stepped down slowly and moved softly up the path,
all the while filming. At one point, we lost sight of them behind the
undergrowth and went further on to the end of our path. As we stepped
around the corner, one of the leopards stepped directly on to the path in

front of us and stood on the slope. I will never forget just standing there, barely ten feet from magnificence. Then slowly he moved away, slinking back into the undergrowth to join his brother and my last glimpse of them was when they climbed the slope and vanished behind some trees. At no point was there any aggression or even any threat. These were cats assured in their surroundings, and from the healthy look of their coats and paws, in their prime and well fed. One of them was just curious and bold and decided on a closer look at us. Since we were calm, quiet and in a group and had the camera, we did not feel vulnerable either. We made no move to go closer nor did we run away or make any sudden movements. Curiosity satisfied, he moved on. Since then, I have seen several leopards in the wild, both in India and Africa, and they have always kept a distance and vanished, unless they were up in the trees. Clearly, this particular leopard was less fearful and more curious, but even he left in under a minute. His brother never came out of the bushes. This is the normal leopard behaviour I am used to and know about.

But something was changing and not for the better.

It was a cold dark night in the hills of Pauri Garhwal in 2005. Gargi and I were walking with two research assistants into a forest area that adjoined many villages. There had been several cases of leopard sightings here with the big cats killing not just livestock but people as well. Between 1995 to 2005, 300 people had been killed by leopards in these hills. Two days before we got to the area, a little girl had been killed. We were there to meet a team of researchers who were studying leopard behaviour, ecology and conflict. They were also doing a deep analysis on leopard scat to see if there was a DNA link to all the leopards killing people. The big question was, were the leopards of Pauri killing people because of injury and inability to hunt their own prey, lack of said prey or were they learning to kill people because they were being taught that people were prey.

All wild animals fear humans. In their deep evolutionary memory it has been passed down that we are the most frightening creatures on the planet. In the early stone age when we had no fire, no shelter, except for caves, and no weapons, we were easy pickings for the big cats. But from the discovery of making our own fire, better shelters and weapons starting with stones, sticks, spears, arrows, and then guns, we quickly became the biggest predators. Animals are clever. Instinct is what keeps them alive and instinct soon started to tell them that we were dangerous and had to be avoided at all cost. It is why the most number of animals will normally be found in areas with least disturbance by people. It's interesting to note that in forests with indigenous people, often, bonds of mutual respect and knowledge kept people and animals on a less confrontational path. Our early tribal people in India would have learnt to avoid predators at all costs, learnt how to behave on seeing one, and would have been armed with poisoned arrows and spears and knives making them formidable opponents. They would have also known exactly when, where and how an animal was moving in the jungles and stayed away from them. It was after the huge population boom in the country and when villages started to spread out and agricultural fields started to take over wild spaces that confrontation began in earnest.

It is when they become habituated and used to us that they lose this fear. This is one of the biggest reasons why captive wild animals can never be released into the wild.

Coming back to the leopards of Pauri, could it be that they had lost this fear and, more alarmingly, could it be that they were learning to kill humans because they had seen it done from when they were cubs? Could their mothers have taught them to do so?

These were fascinating questions I wanted answers to. It is a very well-known fact that man-eaters are made, not born. Usually, an old animal unable to hunt the fleet-footed deer or an injured animal turns to man-eating. Even this is a very rare thing. Most of the time, an animal might kill a person in a conflict situation when it is startled or scared, but nine times out of ten, it will not eat the person, simply leaving after killing him or her.

In 1910, Jim Corbett killed the infamous Panar man-eating leopard reputed to have killed about 400 people. In 1926, he killed the man-eater of Rudraprayag who allegedly killed 125 people. Both leopards started their killing spree after disease outbreaks like cholera and typhoid caused mass deaths and bodies were hurriedly disposed of without proper burials.

Other man-eaters killed over the years across India by hunters have shown signs of old age with no teeth and serious physical injuries. It is very rare that a healthy young animal turns to killing people.

So what was going wrong in Pauri? The leopard that was killed a day after we got there by a professional hunter called in to appease irate villagers certainly had no physical injuries. It was entirely possible that they had killed and shot the wrong leopard. This is where the shooting of so called man-eaters or problem animals becomes tricky. There is no accurate way to tell if the animal that is either trapped or killed is the animal responsible for the attack or kill.

Gargi and I walked to the village where the little girl had been taken. A small hamlet high in the mountains, it was surrounded by forests on all sides. Small patches had been cleared for cultivation and dense undergrowth covered areas around the huts. Two women had been killed while crouching in the dark to go to the toilet and the little girl had been dragged out of her hut. Her devasted father told us that she was asleep near her mother and had been taken away silently. There are no lights here to illuminate the dark and, often, the family slept outside in the fresh crisp air while locking their livestock in. This was something I had seen before and would see over and over as I travelled across India trying to understand man-animal conflict. People either sleeping out in the dark while their animals were locked up, or sleeping in the hut with their dogs and young goats. Leopards come in using the dark and the vegetation as cover, looking for livestock or dogs, and then end up killing a child. Usually this first kill is opportunistic or accidental. Sometimes, after one kill the animal stops but at other times realising the ease of the kill, it can kill again and again.

It was heartbreaking to talk to the father who wanted all the leopards in and around his village to be either killed or taken away. Other villagers, too, were angry because they had lost their livestock to the leopard. Only one old man said to me how leopards had always been in these hills and walked side by side in silence and secret, never showing themselves until these last few years, when they were coming into the village a lot. 'The forest is angry,' he said, 'we never had these problems. Maybe a cow or goat but never a person.' I asked him about the legendary man-eaters and how these hills have become a man-eating hot spot in the last decade. He said that people had forgotten what it was to live with these animals side by side. 'I used to walk everywhere and I had no fear. Sometimes, I would look up and there he would be in the tree, just lying, and I would walk on. Now I fear. The spirits are angry. I don't go out in the dark.'

Two indisputable facts that were proven by the research team were: One, the leopard's prey base. which was the wild boar, the ghorals and the deer, had practically vanished from most of these forests and two, between hydroelectric projects, the spread of towns and a growing human population, the leopard's natural habitat too was in decline, forcing the cat into contact with people.

The reason the leopard's prey base had done a vanishing act was not just because of this decrease in the habitat, but also because of local poaching.

Again, as I travelled across India, these two factors stared me in the face everywhere.

However, the research could never conclusively prove that these leopards were learning to become man-eaters. But other facts proved attacks were higher in areas with no electricity, young children were the main victims and more women were killed than men. Attacks took place mainly at dusk when children would come home from school or play and women would return from fields or when they would go out to relieve themselves. Attacks also occurred at night and usually were higher in homes that had cattle.

Over 150 leopards have also died in this conflict.

Leopards are the most elusive of all the big cats in India and being predominantly nocturnal, they are hard to study in the wild. Unlike with tigers, there is no hard or accurate estimate on leopard numbers in the country. All numbers are a guesstimate. The common leopard is found across the country in nearly all ecosystems except higher up in the Himalayas where the snow leopards are found. They are also highly adaptable cats. Unlike the tiger and the lion, their relatively smaller size enables them to move undetected in depleted forest covers and scrub jungles. Being somewhat arboreal, they also move through the trees high above us. Their greatest survival mechanism is that they are generalists when it comes to feeding. They do not always need the large prey sizes that tigers and lions do. Leopards have been seen to survive on small rodents in many parts of India. Goats, dogs and cattle keep them well fed too.

Two simple solutions – a light outside a home and no weeds creating dense covers in areas around huts – immediately stop many attacks. Separating livestock from the human living area also reduces the loss of human lives.

Of the close to 70 million dollars that is spent on conservation management in India, much of it is directed towards the tiger. This skewed ratio has its own problems. While tiger forests do account for magnificent biodiversity, the very presence of tigers pushes the leopard to the periphery of these forests. Leopard cubs will be killed by tigers and a large male tiger is not accommodating to a leopard in his territory. With rampant development also bifurcating and destroying remaining habitats, buffers, smaller forest patches and corridors, the leopard really has nowhere to go.

And this was something I learnt after 2004, in Mumbai city.

It was a muggy December day. I was sitting inside a cage. I was doing a PTC, piece to camera, explaining how a cage trap for leopards worked. The cage was divided into two sections. In the one smaller division was a dog. The larger area was divided from the smaller area by bars set very close together. The opening to this larger area was weighed down by a weight and a counter weight. The idea was that the leopard would be enticed to come into the cage because of the dog. Once the animal stepped in, his or her weight would cause the counter weight to become lighter and it would shoot up slamming down the gate to the cage, in the process, trapping the leopard. Sure enough, when I crawled into the cage, the gate fell slamming shut in under a second. I weigh about 58 kilos and the average leopard weighs about as much. A big male could weigh even more. These traps were being set up in and around the Sanjay Gandhi National Park (SGNP) in the heart of Mumbai city. SGNP serves as the lungs of this megapolis. Without this park, Mumbai will choke and die an inglorious death. On an average in a single day, the air is so bad in Mumbai that it is like smoking two packets of cigarettes. Sanjay Gandhi is also a catchment area feeding the rivers and streams that flow through this city replenishing the fast vanishing ground water. Unfortunately, it is also a park that covers a massive area of real estate gold. In a 103 sq km area to be precise. It has been targeted over the years by governments and private companies and the only thing keeping this park from vanishing into thin air like so many other forests in the country are the leopards. This is a national park meant for them.

While leopards are only listed as vulnerable by the International Union for Conservation of Nature (IUCN), in India they are a Schedule 1 species. This means they are seen as highly endangered and killing them is a non-bailable punishable offence. This, however, has not prevented the poaching of almost 4,000 leopards from 1994 to 2013, numbers collated by the Wildlife Protection Society of India (WPSI). These are just figures from reported seizures and carcasses. The actual number could be much higher. It is also guesstimated that there are about 12,000 leopards in the wild in India. Camera traps set up around villages near leopard habitats have captured photographs in the hundreds. Each leopard can be identified by his or her set of spots. None of them are spotted alike. This has actually proved that there are more leopards living on the periphery of forests and near villages than in the forests themselves. It's an alarming trend indicating the growing human-leopard conflict.

In Mumbai, in 2004, for two months before we got there, twenty people had been attacked by leopards. Several of them had even been killed. This was not a situation where the animals were man-eaters but one where people and leopards were coming in contact regularly with dangerous results. There had been rioting and the people were demanding that the leopards be all removed or killed. All the attacks were in settlements near the national park. This was a cauldron bubbling to boiling point in the last few years.

Hundreds of illegal encroachments have crept nearly to the Park boundaries. On one side are the huts and slums and on the other, luxury buildings and high rises in the 'no development' zone. Many of the illegal slums that have cropped up house people sent in to squat by prominent builders in the city who later hope to cash in when the area becomes regularised. Big advertising hoardings say, 'Come live with fresh air and greenery', enticing people to buy homes here. What they don't say is 'Come live with fresh air, greenery and leopards'. The leopards come as a huge shock to the residents when they do finally see them. For years, people living in and around this area have seen leopards walk by late at night or very early in the morning and at a distance. In those two months in 2004, however, the encounters had

A rescued leopard. As usual there is a mob watching. The biggest problem is not the animal but the crowd and their behavior. At least this leopard was rescued before being beaten to death as many others are.
Photo credit: Kalyan Verma

been too close for comfort. The national park even has the Mumbai film city encroaching on its premises. This particular spate of attacks seemed to be mostly by female leopards hunting for food to feed their cubs. Several of the attacks were also inside the national park. Here, too, the leopard's main prey base, the wild boar, wild hare, deer, have vanished leaving the forty odd leopards in the park looking for food outside. Much of the leopard's prey has vanished due to local poaching. Catching wild boar and deer is easy meat for people who otherwise starve for a lack of livelihood. With young children to feed and no job prospects it is hard to assign blame. The reality, though, is a situation of severe conflict with no immediate solution in sight.

The forest department was busy trapping leopards and moving them out of Borivali. Nearly twenty leopards were trapped and moved in the course of that year. The human settlement had brought along with it garbage, rats, dogs, chicken, goats and cows. These are by far easier animals for the leopard to catch. As Bittu Sahgal, conservationist and editor, *Sanctuary Asia*, told me, 'Dogs are to leopards what chocolates are to children.' This proved to be true. In three of the attacks, the leopard had gone for the dog and ended up mauling people.

The trapped leopards were a piteous sight. Cramped in small steel cages, the animals repeatedly hurled themselves against the iron bars, injuring themselves and bleeding copiously from gashes on their heads and noses. Barely able to stand up in those cages, they were crazy with despair. It was one of the worst things I have experienced and I will never forget the look in their eyes. A combination of fierce wildness and terror. Many of them stayed in those cages for a long time while some were moved to zoos and rescue centres run by the government.

There is absolutely nothing in the world quite as horrible as the sight of a wild spirit in a cage. All these leopards were adults and sub-adults who had been born wild and born free and now would stay trapped for life. Many of them would become broken and insane from the experience. I have seen them gnaw on their own paws and bite their tails and hit the bars of their cages again and again and again. It makes you want to crawl out of your skin and scream.

I really wondered why we did not just put a bullet in their brains and put them out of their misery.

Many of the trapped leopards were not even animals who had attacked people, they were just animals who were in the wrong place at the wrong time, making the whole situation even worse. Three years later, I visited some of these rescue centres that housed the trapped animals and they were still filled with rage and despair. The prolonged captivity and feeding had not tamed them even by a hair. They were still slamming their bodies into the bars with rage when they saw people and the growls and roars were as ferocious.

Everyone was screaming about how there were too many leopards in Borivali while Mumbai had a population of 16 million and growing. The air was thick with irony. On just one train crossing in Powai, dozens of people were killed every year. No one was demanding that the trains be stopped. These leopards were not transplanted into Borivali after the people moved there but had been there for much longer than Mumbai was a city.

It boggled my mind that people would move into an area with these magnificent wild predators and then be angry that they were there. While there are adivasis here who have been living here for generations, most people here are transplants.

Even here, prior to these few months of conflict, for years things had been calm when there was far less disturbance.

Now the leopards were feeling squeezed out and their normal prey base was diminishing and the chicken, dogs and goats were easy pickings.

While leopards are found everywhere except in the mangrove forests of Orissa and West Bengal, there seem to be certain hot spot areas of attacks. In recent years, most attacks seemed centred around areas that have started growing sugarcane crops. In Gujarat, Maharashtra and Karnataka, during the harvesting season of sugarcane, several leopard cubs are found in the fields. Sugarcane provides a thick cover for these animals who have lost their forests to this crop. Being a water intensive crop, there is plenty of water for them to drink. People who look after the crop live there with their livestock providing them with food. Often, attacks occurred when people would step into the sugarcane and startle a mother with cubs or when a hungry mother went after livestock to feed her cubs.

In Gujarat, a few years before, in the Gir Sanctuary, we joined the forest department as they tried to remove a leopard from a well. The open well luckily had some water in it, thereby preventing the cat from breaking its bones when it fell in. Hundreds of open wells dot the buffer areas of many of our forests and hundreds of animals fall in and die. This particular leopard was lucky enough to be rescued. But the most ridiculous thing was when word got out that a leopard had fallen into a well, hundreds of people came out of their homes, including children and sat around watching the rescue like it was a spectator sport. The mob was loud and agitated the animal further. The rescue operation was tedious and fraught. The noise aggravated the tired and stressed cat who was churning about in the water. The forest department took several hours to figure out how they would get the leopard out. To be fair, they could not tranquilise him as he would have drowned. So now they were faced with dragging a huge male leopard out of a well while he was spitting mad and awake. A huge iron cage was dragged to the edge of the well. It has a sliding door that rose up and dropped down. Three men were positioned on top of that cage with ropes holding the gate up. A pulley system was rigged to the edge of the top bar of the cage. Through this pulley a large rope was fashioned like a noose. Then the department had to drop a thick plank of wood across the well so that two of the guards could sit on the plank centered in the middle of the well in order to drop the rope noose to lasso the leopard. This seemed pretty unbelievable to Gargi and me. We could not see how they would get the rope around the leopard. They could not afford to lasso it by the neck as that would strangle it. They could not lasso a paw either as that would break the leg. They had to get the rope around the middle of the cat. This meant dangling the rope on the water, somehow getting it over the head and forelegs of the animals and then around its tummy. All this while the animal was floundering in the water agitated by a howling crowd. Somehow after hours of trying they managed this miracle. I must say that the department tried their best. It was a murderously hot day and stress levels amongst the humans too was high. They knew that the leopard too would start to flag and if he got too tired he would drown. The minute that rope went around the animal's waist he was just yanked up by a dropping counter weight and swung into the cage and the men quickly dropped the cage door. Now they had to stick their hands in and quickly cut the rope of the animal. The animal was too tired to do much but it was still formidable with its slashing claws and snarls. Then they did the most astounding thing. They loaded the animal onto a truck and paraded it around the few villages in the area so all the people could see it before it was taken to the vet. At the vets, it was finally tranquilised, examined and microchipped. It was released into the forest two days later free of injuries.

Two years later, I would see an angry mob beat a leopard to death in some footage sent to us by the local stringer. The leopard would also maul two people before it died. This was in Meerut. It was about 8 in the morning and a leopard had been spotted in the parking lot of a new housing colony that had come up in an area that used to be a scrub jungle. The minute word got out that a leopard had been sighted, people spilt out of their homes in droves. The noisy and frenetic mob send the animal into a complete frenzy of fear and it ran around trying to escape. However, the crowd was so large that everywhere it turned, its route was blocked. In desperation, the animal even tried to leap up into a balcony and there, too, there were people. In its frantic scramble to escape, it mauled people in its path. This made the mob violent and the next few minutes were horrifying. The forest department who had reached there with a tranquilising gun had no hope of controlling this crowd. The police were part of the mob and there was utter chaos. The mob picked up bricks and other construction material and the animal was beaten to death. I just could not understand what the people were thinking. Why would anybody come out of their house on to the street when they hear a leopard is on the loose? How is this allowed? I asked the police later why they had not controlled the mob and they said that the people were not doing anything illegal by coming out of their homes. That was also true.

Little did I know then that over the next ten years, I would see dozens of such incidents play out. The mobs are still out of control. Most of the time people are injured, there are stampedes and the animals end up dead. There seems to be no co-ordination between the forest department, the police and the fire department. There is no crowd control practised and the situation just deteriorates into total chaos. I asked some people in the crowd why on earth they thought it was ok or even safe to come out of their homes and I was told that they wanted to see the leopard. Did they think it would just walk away quietly or sit still? Did they think that all the noise and chaos would make it practically impossible for the animal to stay calm and for the forest department to tranquilise it? They had no answers to that. It was a spectator sport they wanted a part in.

I have watched a father weep after having a leopard take his son when they were standing side by side after dinner and he did not even see the animal creep up to them. I have also held a young leopard with severe cerebral palsy having been birthed violently after her mother was beaten to death in a sugarcane field. There are no comfortable answers to this mess. As a journalist, that was what frustrated me the most – that I could not wrap this story up with neat answers.

To understand some of it, I spoke to Dr Vidya Athreya, whose extensive study of this conflict threw up certain interesting truths. The main one was that tranquilising and moving the 'problem' leopard was no solution. The leopard, like most big cats, is fiercely territorial and when moved, makes a beeline back to his or her own habitat. They have a strong homing beacon that will, if not disturbed, take them back to exactly where they were taken from. However, this animal that is now on its way back home is stressed, fearful and unsure of its territory, making it a dangerous animal. In fact, in 2001, after the first few attacks in Junnar, close to 100 leopards were reportedly removed and translocated and yet the attacks continued unabated, almost increasing in intensity. Dr Athreya proved through her research that conflict increased in places where suddenly there were more leopards than there should be because of this translocation. She also pointed out that new leopads that moved in after the resident leopard was removed then proceeded to be a bigger problem. Dr Athreya with her years of experience, said that in places where people were tolerant of their presence and accepted that leopards lived amongst them, conflict was minimal. In places where they were not used to the animal and suddenly had one appear due to relocation conflict escalated. Even in Africa, all attempts at translocating leopards have been a huge failure. It just means the problem has been moved to another area.

I was in the Cederberg mountains where the Cape Leopard Trust was doing research on the cape mountain leopards and trying to save them by mitigating conflict with farmers. The cape mountain leopard is much smaller than the common leopard found in Africa. The males weigh in at 35 kilos while the females only at 22 kilos. The average leopard in Africa weighs twice as much. Like our leopards they are solitary and have large territories. Often it is thought there are several leopards in one area when they might be only one or two that are spotted over large distances as they move around. This is why camera traps are of vital importance as they prove that it is the same leopard being seen everywhere. If tigers are identified by their individual stripes, leopards are identified by their individual spot patterns. Just like with tigers, a male's home range is larger than a female's, and several females can be found within the home range of a male leopard while he will not tolerate the presence of another male. It is another guesstimate that there are only about 1,000 cape mountain leopards. Cub mortality is high and many leopards are trapped and killed. Here in the Cederberg, leopards were dying because they were being shot for predating on the livestock. Some deaths were unbelievably cruel with gin traps being set up to catch them. A gin trap is a steel trap that clamps around the animal's paw when it steps in on it. It is a long horrible death filled with unbearable agony. Even if found, usually the paw needs

Adult tiger resting in a bamboo thicket.
Photo credit: Kalyan Verma

Two sub-adult tigers in Umred Karhandla Tiger Reserve.
The brothers, sons of the legendary tiger Jai, are the future of this relatively new tiger park.
Photo credit: Tom Foster

My favourite lion. Tsau. A white lion. He was rescued from a breeder who would have sold him into canned lion hunting. I have known him since he was five months old.
Photo credit: Jurg Olsen

Tequila, the jaguar. I have known her since she was a cub. At the Jukani Predator Park, run by my friends.
Photo credit: Jurg Olsen

The wild cheetah who allowed me to walk with her for over two hours.
Photo credit: Craig Foster

Standing holding an elephant leg bone in the Savuti. Victim of a lion kill.
Photo credit: Craig Foster

Male lion and big male elephant face off in Savuti.
The lions will not attempt to kill such a large elephant unless desperate for food.
Photo credit: Swati Thiyagarajan

Mother with cub. Asiatic lions in Gir.
Photo credit: Dr Anish Andheria

Leopard in the wild in India. Leopards are elusive cats, hard to spot,
even though they live closer to humans than the other wild cats.
Photo credit: Kalyan Verma

A wild leopard that has been caught and caged after wandering into a densely populated area.
She might have to spend the rest of her life in a cage.
Photo credit: Kalyan Verma

The classic leopard sighting. Sitting on a tree.
Photo credit: Kalyan Verma

A young leopard sharpening its claws on a log.
Photo credit: Kalyan Verma

Female Asiatic lion resting in the Gir forests.
Photo credit: Dr Anish Andheria

Shetty and me enjoying a laugh when I was trying to do my piece to camera.
Photo credit: Craig Foster

A wild, mother sloth bear, with her cubs. The cubs ride on their mother for a while until they are much bigger. When the cubs are caught in the wild, their mother is invariably killed.
Photo credit: Kalyan Verma

Domesticated African elephants, tamed for tourists in South Africa. They have killed several people, mostly their trainers.
Photo credit: Craig Foster

An Asiatic elephant mother and calf crossing the road. Most elephant corridors in India are ruined by roads, dams and other development projects and this has become a huge problem for herds.
Photo credit: Kalyan Verma

An Asiatic elephant playing in the water.
Photo credit: Kalyan Verma

The large herds of African elephants were the highlight of my trip to Kwai Conservancy in Botswana.
Photo credit: Craig Foster

Having breakfast near my tent with a large male African elephant having his
on the other side of the tree. In Savuti, Botswana.
Photo credit: Craig Foster

Elephants at the water channel in front of my camping site in Savuti.
Photo credit: Craig Foster

An African penguin who has just come out of the water.
Photo credit: Craig Foster

A nesting tree in the village with loud and hungry pelican chicks.
Photo credit: Kalyan Verma

A pelican flying with nest building or repairing material in Kokkare Bellur.
Photo credit: Kalyan Verma

A mermaid's purse or a shark egg.
The shark egg is coiled around a Sargasso frond and the young shark will emerge from it when its ready.
Photo credit: Craig Foster

A great white shark cruising past in False Bay.
Photo credit: Craig Foster

A puffadder shyshark. One of the smaller sharks found in the kelp forests on the coasts of the Western Cape.
Photo credit: Craig Foster

A dark shyshark.
Photo credit: Craig Foster

to be amputated and the animal has to live in captivity all its life. Unlike India, farmers in South Africa, Namibia, Zambia, Zimbabwe and Mozambique can shoot leopards and get a license after the fact if they can prove that the animal was a threat to their livestock. Unless the leopard is updated to the endangered status, farmers will always be allowed to shoot it.

Here, the Cape Leopard Trust is trying to create a concept called 'predator-friendly meat'. This is so that consumers of meat know that the meat they are buying did not involve the deaths of leopards and other cats and animals. In order to do this, they are trying to increase the farmers' awareness of leopard behaviour and involve them in the conservation of the leopard. Two preventives are being tried out. One, large dogs like the Anatolian shepherd dogs are being given to farmers. The Anatolians are large dogs that are used to ward off bears in the mountains of Turkey. These dogs are not pets. They are housed with the livestock from when they are puppies and live with them. The herd becomes their pack and they will stop at nothing to protect the herd. These dogs now being used to protect livestock from leopards and cheetahs are helping greatly with livestock predation. The other preventive is flashing solar collars around the sheep and cows. These flashing collars have small lights that blink like Christmas lights in the dark and throw off a predator looking down on a herd with an intent to kill. A few national parks in India are now trying these solar flashing lights to fence in crops to keep deer out, and are succeeding. Most big cats have very sensitive eyes and can see best in low light and in the dark. Bright lights in their face disconcert them and flashing lights confuse them. That is why even a simple torch, when it flashes bright, keeps big animals like lions at bay. Slowly, progress is being made in the Cederberg and the farmers are now making more money on their predator-friendly meat as consumers like the idea that they are eating something that did not cause the death of a beautiful wild animal.

Even in India, the solution seems to lie in making people aware of leopard behaviour, their own contribution to the problem with the sugarcane and the domestic livestock. It seems that solar fencing, solar lights on streets, better cattle enclosures and general awareness when moving about reduces the conflict drastically. In Tadoba, the few villages that have solar street lamps, solar lights in their homes and solar fencing around in their fields and cattle, gas for cooking and access to toilets have reported a drastic drop in conflict situations.

The one other measure is cattle compensation. Paying compensation for killed cattle goes a long way in assuaging people's anger. Dozens of leopards have been killed after the cow or goat that they have killed has been poisoned by the villagers. It is usually rat poison and it is a horrible painful death for the animal. If the leopard is a mother, her cubs will die without her or die when they too eat the poisoned meat. If it is a male leopard, his death leaves a void that will be filled by other lesser predators like caracals or jackals or even more leopards who might not be as strong as him, so that a territory ruled by one big strong male in his prime could be taken over by two or more lesser males. Camera trap footage of carcasses show that often, it's not just the leopard but, sloth bears, hyenas, jackals and vultures feeding on the one carcass. If it is poisoned then, it is death to all those animals.

Recently, as a first in India a leopard conservancy has been started in Jawai in Rajasthan. It has been set up in an area with a high leopard population. The concept is to bring in tourists who want to see leopards and involve the local community in the operation of this conservancy. Slowly, the local people are beginning to realise that more leopards means more tourists and that means more money. Jawai however is unique in that, the leopards here are almost twice the size of the average Indian leopard with some males reaching a size of 120 kilos. The traditional relationship between the people and the big cats is also very strong, with the leopards not ever having attacked a human here. People here worship the animal as a god, tolerate the loss of their livestock and have been living in harmony with them for generations. This gives credence to the fact that a general understanding and acceptance leads to greater co-habitation of people and predator.

This conflict might seem like something that is happening far away from us, well removed from our safe homes in the cities, but as Mumbai proved, that is not true. In the last few years, several towns not far from major cities have had leopards come through their streets. I would like to also mention here that every time we drink tea or coffee, we should know that nearly every tea and coffee estate in this country is in leopard territory. Every spoon of sugar comes at the expanse of vanishing leopard territory. Every highway we enjoy, every drop of water comes from hundreds of kilometres of canals that cut through forests. Even the food we eat is from fields that have taken over forests while all that electricity is from coal fields that have swallowed leopard land. When a leopard attacks a child in a far away village, it is happening to all of us. It's just that we enjoy the luxury at a price someone else and the leopard is paying. In Africa, several countries allow the trophy hunting of leopards. This is a problem as with little knowledge of real numbers on the ground it seems strange that hunting quotas can be set. There are also no rules in place for whether only older non-breeding animals can be hunted. As leopards are not group animals and are hard to track follow and study, it is hard to figure if the animal is older and unable to breed. So most of the hunted animals are potential breeders and this is a huge problem for the future of leopard populations. It is also hard to tell if the male leopard killed is a father with young cubs. Such a kill can potentially threaten the cubs as a new male in the territory will kill them. South Africa banned the killing of leopards in 2016. This comes after several scientists and conservationists sounded a warning about the leopard populations. According to CITES rules, South Africa is allowed a quota of 150 leopards a year for hunters. Leopards are also bred in captivity for canned hunting just like lions. Infact in Africa, many canned hunting farms have nearly every type of wild cat on tap for hunting.

And if the only leopards I ever knew about were the ones in conflict, it might have all turned out differently for me in how I understood them. But then I met Baby.

Gajendra Singh and Visalakshi Devi were handed over two young leopard cubs in 1997. A farmer had found the cubs in his field. Their mother stashed them in the field for safekeeping and must have gone out for a hunt but the farmer, assuming they were abandoned, picked up the cubs and brought them to the forest department. Instead of sending them off to a zoo which was overcrowded with leopards anyway, the forest department asked Gajendra Singh and Visalakshi Devi, a couple who ran a resort in Bandipur, if they would look after the cubs until a solution could be found. They said yes. The cubs, a male and a female, were named Baby and Bully, and they had barely opened their eyes. She had to feed them with a bottle and then hand reared them. The forest department at the time decided that they were going to try something new. They were going to try and raise these cubs in semi-captivity and then release them into the wild.

Baby and Bully were released into a large enclosure at the bottom of their property in Bandipur once they were old enough to eat solids. Within the enclosure, live prey would be released and the cubs would learn to kill them. Once they were sub-adults, they were taken and released into another enclosure in the heart of the forest. They were fitted with radio collars and would be taken out on leashes on regular walks, Bully was feisty and curious and Baby, while curious, was soft and gentle. After a few weeks of these long walks, the leopards were let out without leashes, but they would always come back to their enclosures. After a few weeks of that, Bully started to stay away more and more while Baby would still come back to her 'parents'.

One day, Bully did not come back and his carcass was found by tracking his active collar. He had attempted to kill a sambhar deer and was gored in the process. Even as cubs, while Bully had a rough way of playing, Baby never ever scratched her human family. Bully's death made everyone nervous but they persevered with Baby. Soon, she started to go out and stay away for days. When her human family would go out and call her, she would always come to them. This is something I have seen myself in Bandipur. She would come and back-rub herself against them and walk with them and gently caress them with her tail. Soon, her collar was removed and still she would come if they drove into the forest and called for her. Even though she was hand raised, she never ever went to the other tourists who would come to the park. She would only come to her 'parents'. One day, she vanished and disappeared for much longer than she ever had before. This was when the miracle happened and it forever changed what I thought was my understanding of leopards.

Leopards are one of the trickiest of all big cats to deal with. All big cats are dangerous no matter how long they have been in captivity and how deep their bonds with their handlers might be. But you still see tigers and lions in even stressful captive situations like circuses. You will never see a leopard.

It has always been understood that they are just too dangerous to be around or work with.

Baby came back but was agitated. She seemed restless and kept walking away but looking behind her. Visalakshi Devi suddenly realised that Baby was asking them to follow her. So they walked behind her for a while until she stopped at a really dense thicket and then crawled in and came out and waited for them until Visalakshi crawled in with her. To their delight and shock, they found that Baby had cubs. She wanted to show her human family her cubs. Baby continued to bring her cubs even when they were older, to her human family but the cubs would always stay away being wild leopards with wild protective instincts.

She has even allowed her 'father' to film her on a kill. All predator knowledge tells you that adult big cats are most dangerous in the moment of the hunt when they make a kill. This is usually when they can attack anything around them. Baby, however, holds the cheetal deer down by the neck while calling for her cubs to come finish the kill and Gajendra Singh filmed her with the live deer clutched in her jaws. After getting the shot he wanted, he left, knowing that the cubs would only come out after he left.

Baby is the only story of rehabilitation of a hand-raised leopard in the wild. But what was more miraculous was how she continued to care for and love her human family even when, for all intents and purposes, she was living as a wild leopard.

One day, as all leopards and indeed any living thing does, she died and her family knew she had when she never came to answer their calls. But in her lifetime, Baby taught me that assuming knowledge of any animal just as a member of its species was wrong and that every animal was an individual with an individual's responses to a situation.

To me, more than tigers and lions, leopards really signify the magic of the word *wild*. There is something in them that is raw and fierce. In a raw comparison of capabilities, for their size what they can do is more powerful than what lions and tigers can with their strength. In that scale of comparison, they are actually the most formidable of all the big cats. And yet a leopard like Baby makes you think there is so much we just do not know about these magnificent animals. For every ten people studying tigers, there might be one studying leopards. Elusive secretive and nocturnal, they are the true spirits of the forests—owning both the ground and the trees. They might not be classified as endangered yet but they are getting there. The loss of the leopard will be the loss of all that is wild in soul and spirit.

> *Those who have never seen a leopard under favourable conditions in his natural surroundings can have no conception of the grace of movement, and beauty of colouring, of this the most graceful and the most beautiful of all animals in our Indian jungles.*
> – Jim Corbett, *Man-Eaters of Kumaon*

4

SLOTH BEARS
WHAT A DANCING BEAR TAUGHT ME

And maybe... you are a little fat bear cub with no wings, and no feathers.
— Else Holmelund Minarik

THEY WERE so small they fit in the palms of my hands. Their eyes had just opened and they were crying in distress. Every time they felt

my finger in front of their mouth, they started to suck on it. They
were trying to climb up my arms on to my back. The feel of my hair
comforted them. I was in the Agra police station where two tiny sloth
bear cubs had been seized from poachers.

Both cubs only stopped crying when held close to our chests. Gargi
and I did not feel like putting them down. I took comfort in the fact
that they were going to a rescue centre set up by Wildlife SOS India,
co-founded by two wonderful people, Geeta Seshamani and Kartick
Satyanarayan, where they would get lifetime care. I would have taken
greater comfort if they could have been released into the wild when
older, but that I knew was a chance they had lost forever.

When I was seven years old, I found a little bird hopping around
in my garden. I watched it for a while and realised that it was lost
and scared. I felt its frantic worry and confusion. I also realised it was
a baby bird as it could only hop about and cry but not fly. I did not
know what sort of bird it was but I knew one person who would help
it, my uncle Siddharth. He lived about ten minutes from my house. I
knew I had to go into the house and call him but I was worried that
if I left, it might get killed and eaten by the neighbourhood cats or
my dog. So I did what any seven-year-old would do – I called out
for my mother. She came rushing out of the house because I must
have sounded frantic. I explained the situation to her and asked her
if she would watch the bird while I called Siddharth uncle. I ran in
and called him and tried to describe the bird to him. I told him it
was tiny. Walked in strange hops, had a high squeak, which was like
a cheep cheep, and was yellowish green in colour. He immediately
told me that the little bird was a common tailorbird and that she was
probably scared, as she must have fallen out of the nest. He asked me
to keep an eye and ear out for the adult bird, which he said would
be greenish yellow on top with a white underbelly. He said the adult
would have make cheep cheep sound too and would be flying back and
forth looking for the baby.

Uncle assured me that he would come immediately. He was home
in ten minutes. In that time, I looked around for adult tailorbirds and
kept an eye on the baby. When I saw Siddharth uncle, I felt immense
relief. I felt that now everything would be ok because I had seen him
help so many birds over the years. He sat with me and said we would

wait for the parent bird. I asked him why they were not looking for the baby. He explained that the baby must have fallen out of the nest while the parents were out looking for food and that as soon as they returned, they would start looking. He offered to walk around my garden, looking for the nest. He said that the baby bird could not have hopped too far. I had learnt from him that touching and handling birds was a tricky business because it did two things. In the case of an injured bird when handling was a must, it still increased their stress levels. In the case of baby birds, it not only increased their stress levels but also transferred human smells on to the bird, which sometimes made their parents reject them. He also told me to talk to the little bird and tell him that we were helping him and would try and get him home. I did anything uncle Siddharth asked me to, so I kept up a steady stream of chatter with the little one, explaining what we were doing. To this day I try and project thoughts of calm and love when I am around wild animals.

Just as Uncle reached the bottom of my garden, I suddenly heard frantic cheeps from the garden wall and I saw the adult tailorbird. Both of us immediately backed away. Uncle told me that we should just stay quiet and watch. On hearing her parent, the baby bird started a shrill call and hopped about madly. It was just sheer magic to watch the adult fly towards the baby and land next to it. For the next fifteen minutes, I watched the parent coax the little one towards a small leafy bush about five metres from where they were. Tailorbird nests are woven together by using the leaves of the bush they are in. It is often very difficult to spot the nest as it just blends in with the rest of the plant. Usually, one can tell there is a nest in a plant only when one sees the adult birds darting in and out. Now that the parent had arrived and was darting up and down, I knew the nest was somewhere in the bush. Now it seems like the easiest thing is to pick the baby bird up and place her in the nest, but I had also been taught by uncle Siddharth that I could leave traces of my scent on the nest and spook the adult birds enough that they might abandon the nest.

Once the baby got to the bottom of that bush, it was a dilemma as to how to get her in the nest. For almost ten minutes, the parent bird tried to get her to climb the bush. It was either still remnants of panic or she was just too small to hop and climb but the baby just sat beneath the bush and cried. Finally, Uncle gave me a freshly washed hanky and told me to use it pick the bird up gently and put her on the bush. He taught me to approach her really slowly and softly. He also asked me to tell the little bird that I was helping. I also had to tell the mamma bird that I was helping. For the first two minutes, as I got closer, the mamma bird panicked and called and flew away and flew back frantically. I stopped moving and concentrated very hard on trying to tell them slowly how I was helping. At the time, I was not to know that what I was doing was an instinctive interspecies communication, but later in life, I would understand it better and even make a movie on the phenomenon that would be watched by over three million people. All I knew then was that uncle Siddharth always said that animals could pick up on our energy and that the calmer and more full of love we were, the more they trusted us. I very quickly picked up the baby and placed her on a branch near her nest, which I could now see. I also quickly backed away. Mamma bird landed near her and nudged her all the way into the nest.

Over the next few weeks, I watched her slowly learn to fly and leave the nest, but thirty-five years later, I still remember clearly the utter joy of knowing she was wild and free. To me, that moment has stayed in my heart, and since then I have been part of rescuing and returning many birds and animals to the wild. But I have also watched hundreds if not thousands of them lose their freedom and endure captivity. It's not always bad if the animals and birds get to go to places where they are looked after well and have their own space, but too often, that is not the case.

This has made me a strong anti zoo person in philosophy, but realistically, I support places that are true rescue centres and sanctuaries because often the animals cannot be released due to injury or are just too young to survive in the wild on their own like the two baby sloth bears.

These two babies were rescued from a poacher who would have sold them on to a community called the Kalanders who use them as dancing bears in India.

A few years before this day, I was in Agra to do a story on the dancing bears. As I drove on the highway towards Fatehpur Sikri, I saw dozens of bears on the side of the road. I was astonished to see this in the open as ostensibly owning and displaying this animal was against the law. On that hot dusty day, as I got out of my car and started to film the dancing bear, I knew I had to make this a strong news story. I had to ask why a Schedule 1 animal could be paraded on a popular highway so blatantly and not have anybody do something about it. The tiger is a Schedule 1 animal too. It would be absurd to watch a dancing tiger on the side of the road, it would never happen. So why was nothing been done about the bears? The bear was standing on its hind legs and swaying from side to side as a drum was being beaten. Have you seen a sloth bear? They are quite peculiar-looking in that their bodies are a dense mass of coarse fur, usually black in colour with some brown and reddish highlights. They have pale muzzles and their lower jaw juts out further than their upper jaw. Their lips seem fleshy and they have long claws on their paws. When they are on all fours, a fully-grown male adult can stand at about 85 cm at the shoulder. That is nearly 3 ft tall at the shoulder. When they rise up on their hind legs they can come close to 5 ft sometimes and a band of light yellow hair in a U-shape is visible on their chests. A sloth bear is an intimidating and dramatic sight and this dancing boy was no different. On closer inspection, however, I could see that he had no claws, his mouth looked like it had no teeth. Sloth bears possess the same number of teeth as other bears (6 upper and 6 lower incisors, 2 upper and 2 lower canines, 8 upper and 8 lower premolars, and 4 molars in the upper jaw and 6 in the lower). Clearly this one had had them all extracted.

Then I saw the real horror. A large thick rope went in through the nose and upper palate of the bear's mouth. The rope was then looped through an iron ring, the other end of which was in the handler's hands. The bear's every movement to the beat of the drums was governed by the jerks of the rope. I was to learn later and see for myself that when they are still young, a red hot needle is used to punch this hole through their nose and upper palate and it never quite heals as the rope constantly chafes the wound. This makes the wound sensitive and every jerk on the rope causes immense pain. A jerk to the left and they move their head to the left, because if they don't, it will be too agonizing. It was also a way the handlers could control the animals even when they were not dancing as adult sloth bears are super strong and can rip into a human in minutes.

Sloth bears are solitary but not aggressively territorial. In many forests they are nocturnal, so they move around mainly at night. Mothers with cubs have been observed moving around in the daylight and this might be to avoid predators.

In the wild, a bear lives up to twenty-five years. In these captive conditions, they die in seven to eight years from malnourishment and infected wounds.

A wild sloth bear is mainly an insectivore and a fruitarian. The insects consist mainly of ground-living termites and ants. They use their strong claws to dig deep into the termite mounds and soil to look for them. They like sugar-rich fruit and also climb trees looking for beehives, as they love honey. Fallen ripe fruit are their usual go-to for fruits and they have been observed feeding occasionally on carrion or left over rotted kills of predators.

In captivity, while fruit can be made available to them, it's harder to provide them with termites and ants. It's a major protein requirement for them and this lack of protein in their diets is a huge problem. In the Kalander community, poverty is so prevalent that even fruit is a luxury and often, some milk and maybe bread with the odd banana becomes a staple. This bad diet combined with their wounds and stress kills them young.

More cubs are then needed to take their place creating a never ending loop of poaching in the wild. The animals are slow breeders and do not breed well, if at all, in captivity. Sloth bears are predominantly nocturnal but mothers with young cubs might move around during the day.

In order for them to be effective as dancing bears, they need to be trained and handled from a young age on. To capture young wild sloth bears, their mother has to be killed. Baby bears do not go anywhere without their mothers. In the wild, cubs are born in secure dens. These dens are either natural caves or dug out by the mother bears. The mother bear stays in the den for close to eight weeks without leaving her babies. At this time, she survives off her own fat and water reserves. Baby bears are born blind and cannot see until they are about four weeks old.

To get to the cubs, the poachers often have to kill the mothers. Cub mortality is very high. They need the powerful nutrients in their mother's milk to survive and need her guidance to reach maturity. A sloth bear mother will carry her cubs on her back until they are almost ten months old and protect them by fighting to the death with any threats. A female sloth bear with cubs has been observed fighting and standing down two tigers when confronted. So when the poachers take the cubs they are also taking out a viable breeding female, endangering the future survival of these animals.

Adult bears are not spared either. Bears are often killed for their gall bladder which is used in Chinese traditional medicines. Many bears also get captured and used as live bile producers. Black bears and brown bears are also killed or captured for the same reason. The bears are trapped and confined in small cramped cages which force them to stay in a supine position. An incision is made into the side where the gall bladder is and the bile is extracted through a tube. It's a horrific barbaric practice that exists to this day even though there has been strong international condemnation.

That day, on the hot highway, watching the dancing bear, I felt anger and a deep sense of outrage. I wanted the bears seized, I wanted the men arrested and I wanted them to pay for what I saw as extreme cruelty. Two years later, I would have a very different view.

My news story was made into a headline and for a while, I heard about that part of the highway being better policed and I felt like I had done something positive. Several bears had been seized and their handlers arrested. I felt like I had contributed to something positive.

This, however, was no solution to the problem. Years later, in the summer of 2004, when Gargi and I decided to do an entire episode for *Born Wild* on sloth bears, I would learn so much more.

I had met Geeta Seshamani years ago, while still a student. She is the founder of Friendicoes, an animal shelter that caters for stray animals. I had often taken many injured stray animals from the streets of Delhi to them and Friendicoes always treated them with great compassion and love. Geeta was a college professor and has a deep and compassionate love for animals, which she channelled into a life filled with working for them. She has a partner, her nephew, a young man Kartick, whom I first met on a snake rescue story. Built large and with a booming laugh, Kartick also had great compassion and passion for animals. His was more of a crusade to rescue animals in distress and

A sloth bear feeding at Dharoji National Park.
Photo credit: Kalyan Verma

stop poaching. By 2004, they had co-founded the Agra Bear Rescue
Centre and were the heart of the Indian chapter of Wildlife SOS, an
international animal rescue and conservation organisation. Over the
years, I had become better acquainted with Kartick and when I heard
of the work he was doing with sloth bears, I decided to feature him in
Born Wild.

It was then that I met a Kalander family and spent time with them.
As part of the story, I wanted to get an understanding of why the
Kalanders were engaged in such a cruel livelihood. My day with the
family turned many things for me on its head. I can't say that this was
the first time my line of thinking changed profoundly but I can say
that this was definitely one of the times I remember feeling a whole
seismic shift in the way I would start viewing people in this landscape
of conservation. The family was an extended one consisting of an old
grandfather, an aunt and uncle and their children, and a mother and
father and their children. They were about twelve members in all. They
owned two sloth bears. There was also a large horned owl. The bears
were tied up in the courtyard and the family sat outside on charpoys
in front of their small mud dwelling. This family had volunteered to
give up their bears to Wildlife SOS. In return, they would get a small

stipend and a loan to start a new life. The father was being set up with a small shop and the women were learning to sew. Small donations were also being sourced to send the children to school. It was only then that I saw the human faces behind the people I only thought of as cruel bear oppressors. I had never once wondered, why would people want to be in this business?

Kalanders are a community of wandering nomads who travel from place to place with their dancing bears, earning money by making them perform. Entire families depended on this meagre income and it was the only way of life they knew that was accessible to them. A community of the minority religion, it was also not easy for them to become assimilated into mainstream life. Their children were never in any one place too long to go to school and often, there was no money to send them either. Also seen as low caste and dirty because of their practice of living with their animals, they were not exactly welcome everywhere they went, even if they wanted to settle down. Some communities had set up small places near Agra and Jaipur. Here, with the highways bringing in tourists, they figured business would at least pay for their food.

Some reports suggest that the Kalander community had started this practice in the thirteenth century in the courts of the Mughal emperors, while other reports suggest that the practice started as early as the Indus valley civilization (7000-300 BC). While the historians might not be able to agree on an actual date, it is very apparent that it is an old and traditional practice. Sloth bears, when young, are very needy. They stay with their mothers until they are almost two-and-a-half years old. They crave contact and need to be held a lot. This makes them animals that are relatively tameable. I say relatively because once they reach sexual maturity, they become very big, strong and dangerous unless handled very carefully. Of course, just like with any other animal or us human animals, different personalities abound, making them more or less tame like the other animals, but for the most part, they stay wild at heart.

This family knew no other way of life. The old grandfather could remember his old grandfather with bears and he grew up with them and that's the way it has always been. Sitting under a tree that day and watching the young children stroke the bear and feed it small bits of roti, I had an epiphany. They were not being cruel. They lived in a world where every meal was a challenge, every day a struggle, and the future was a hazy mirage that actually did not exist. Their lives were cruel. They actually loved their bears. Apart from the horrific nose ring and rope, there were no other marks of daily abuse inflicted on the bears. Yes, some families declawed the bears and removed their teeth, but that was done while they were still young. There was no daily beating, starving or casual cruelty. They could not even understand why I was distressed about the bear's nose. It is what had to be done to control them. The family was scared, puzzled and wary of what their new future would bring.

It struck me then. It is so easy for us in our comfortable lives to freely wave around words like cruelty, abuse and evil, but if our comfortable lives were ripped away and we were left on the streets, what might we do? Yes, poaching was an awful scourge and had to be stopped, but how much of my lifestyle was leading to the sloth bears' habitat being destroyed and them dying because of that? It was an afternoon of uncomfortable questions and intensely uncomfortable answers. When the mother brought me a small cup of tea and causally gave a banana to the bear and I watched the children stare at the food, I wondered, how many evenings had this family gone to bed hungry and yet found some food for their bears?

Kartick and Geeta, via Wildlife SOS, wanted to not just seize bears and arrest people but they wanted the bear handlers to voluntarily hand in their bears and then help them with rehabilitation that would set them on their first steps into a new future.

There are people I have met who find even this shocking. They only feel for the Kalanders and say that this kind of 'do good' rehabilitation does not help in any way except in making us, the elitist conservationists, happy. I have only one counter argument to that. According to an eminent field biologist and conservationist, Dr Yoganand, who has studied sloth bears for years, given the degradation of habitat and the fact that sloth bears are only found in 10 per cent of the available forests of India, there are only about 7,000 of them in India and perhaps between 10,000 and 12,000 in the world. These bears are found in Nepal, Bangladesh, Sri Lanka and Bhutan. Their populations are dropping by the day. Let's say we do nothing, one day there will be no more sloth bears to be had and then where will the Kalanders go? What will they do? By then it will too late to even receive help. Here, at least they were getting something in return for handing in their bears. No one was being arrested and thrown in jail as per the law. That would be no justice nor would it solve the problem.

Five years to that day, the last dancing bear in this area would be rescued and the families helped. If the programme had not been successful, more bears would have not been handed in voluntarily.

In 2008, I would make yet another trek back to the Agra Bear Rescue Centre, this time to look at how successful the programme had been. I would also go back to understand why some of the bears who were rescued as babies could not at least be rehabilitated in the wild.

As I walked in to the centre, a bear who had been handed in just that morning was brought into the hospital. The first thing that happens when the animals arrive is the symbols of its captivity are removed. In this case, the horrible rope and chain. The bear is also examined for diseases and treated. Many bears come in blind, with tooth decay, parasites and extreme psychosis, much like soldiers who suffer from post-traumatic stress syndrome. Some of them also miss their owners. There are many bears here who have been handed in after years in captivity and they stand slightly off and away to the side and seem to sway from side to side to a music only they can hear. Just years of conditioning that are hard to throw away. Over the years, I had been visiting here not just for stories but also to just come and spend time with the bears, and I had gotten to know several of them quite well.

Back to the fellow on the operating table, I watched as the vet examined him thoroughly and then removed that awful rope and chain. This one had hardly any teeth while he still had his claws. He will be isolated for a little while until the doctors are sure he is disease free and healed and then he will be released into an enclosure.

The enclosures here are huge and, depending on the bears' personalities, house either just one bear or several. In the wild, sloth bears are solitary and territorial but can come together to feed although male bears are more aggressively territorial. Cubs and younger bears can be together and once they are together from a young age, they usually adjust quite well together.

I had spent a lot of time with sub-adult bears here and I knew just how sharp the teeth and claws are. Many of the smaller bears often climbed into my lap and gnawed on my hands, my bag and my hair, and climbed my back. They have so many sounds to them. My favourite is the lovely gurgling noise they make when they suck their claws or eat something particularly nice or are generally feeling happy.

It is tricky to rewild them. Many of them are without their teeth and claws, making it impossible for them to go back to the wild. Many are too habituated to people to ever be released as well. Sloth bears are formidable in the wild. They see humans as predators and react accordingly. While in their wild habitat, they mostly stay away from people, in areas where the habitat is degraded and they venture into villages for food, confrontations can turn ugly. Twice, two adult bears have charged me and the experience is not fun. The first time was in Bangalore with Gargi. While we were filming the bears, one of the females suddenly rose up on her hind legs, made a huge huff-like panting sound and then just came at us with her fur bristling. Her mouth was wide open, drool was spilling out and the snarling sounds were intimidating. At a gallop on all fours, they can outrun humans. While Gargi and the others made it to the jeep, I locked my knees and stood my ground.

These are the moments when time slows down and I can see everything in a light golden haze. It's like I am having an out of body experience and watching me from outside of me. I could hear Siddharth uncle's voice firm and loud in my head – 'Don't run. You can't outrun them, so don't try. Stand your ground, send out peaceful thoughts, radiate calm and hope for the best.' I could hear the last words being said with that familiar chuckle of his. I also heard him say, 'Swati, they almost always give you a warning, so watch for it.' Clearly, we had missed whatever warning she had given us. The bear skidded to a stop about a foot from me, went up on her hind legs, snorted once and shambled away. To this day, I don't know if I stayed that still because that is what I know is the best thing to do or if it was sheer terror that kept my knees locked. Somewhere in the back of my head I also remembered that she might be doing a mock charge just to prove she was in control. I was lucky that this was the case.

Sloth bears are not aggressive attackers in that they do not head out seeking confrontation. Usually, the attacks are defensive. Since they are predominantly found in dry deciduous forests that also house the tiger and the leopard, the bears have learnt to deal aggressively with threats. With fragmentation of their existing habitats and loss of forest cover and habitat, the bears are coming into greater and greater conflict with people. In India, it is estimated that there are at least 25-30 encounters with bears every year, leading to maulings and fatalities and dead bears. In Tadoba, near the tiger reserve, villagers told me that they were more nervous about running into the sloth bear than the tiger. My guide, a local boy, told me to come sit in his village if I wanted to see bears every night.

A group of people can easily get together and kill a bear, and one bear, if it has the advantage of surprise, can kill several people. Many bears, not just the sloth bears, but brown bears and black bears have been beaten to death and set on fire, across India, when they have been caught in villages. Several people have also been mauled and killed, making people terrified of the bears.

As I played with the cubs, it seemed hard to imagine that these animals could be aggressive enough to kill but feeling their strength and realising that in the midst of nature, they would hardly be so trusting, I knew that they could be formidable opponents.

In areas like Orissa, Chhattisgarh, Bihar and Madhya Pradesh where mining, dams and deforestation and a growing human population is wiping out forests, confrontation and conflict seems almost inevitable.

Except for Dharoji National Park in Karnataka that has been created for bears, sloth bears today survive in protected areas created for tigers across the country. In Dharoji, the park management actually supplements the bears feeds and does a lot of hands-on management of their landscape that is quite controversial. Many conservationists and biologists feel that this excessive management could affect the way the bears behaved but so far there have been no studies on this and the bear population in the park is doing well. At least as a protected area, the habitat is being protected instead of being taken over for mining like so many other areas in Karnataka.

One of the cubs, a two-year-old named Shetty at the Agra shelter and I formed a nice bond. Every trip I made, I had a lovely time with him and years later, I would go back and have a beautiful moment with him again.

In 2008, I came back to the bear sanctuary, this time with my husband Craig who was my cameraperson. Craig is a filmmaker and one of the best known in southern Africa and internationally for wildlife and anthropology documentaries. I had told him the whole sloth bear story and he was really looking forward to seeing them. We stepped into one of the enclosures to film the bears, and one of them charged us. Kartick was there with us as well and he said 'Don't run'. This time, as I trusted Kartick to know his bears, I stood still without fear. The bear skidded to a halt in front of us after much impressive growling, roaring, snarling and huffing. Then he just moved away. I looked back at Craig who had held still after taking one large step back. 'See I told you', I said, and he laughed. He said that the charge was pretty intimidating.

In the course of twenty years of filmmaking, Craig has been charged by hyenas, cape buffalo, tiger sharks, leopards and the great white shark. So it was impressive that he found the sloth bear charge intimidating. Yes it was, and I can see how things go wrong in a confrontation.

And then up shambled the most beautiful bear in my biased opinion. Shetty. A large male with an enormously shaggy body and mischievous eyes. I wanted to do a piece to camera and Shetty decided to come sit with me while I did it. He sat up against me and grabbed my hand, giving me a bear hug literally and snuffling at my hair. Animals have a very strong sense of smell. While they also have excellent eyesight, it is often smell that helps them identify someone. It was the most fun PTC I have ever done with Shetty trying to put his head under my arm. In that moment, just cuddling with him, I felt my entire heart just open and fill. It is to me, the greatest blessing in the world when an animal trusts you enough to let you into his or her space. Especially when they have no reason to do so or when there is no reward involved. I have had the privilege to experience it many times and just for these moments in my life, I believe I have lived the most blessed life.

There is a look in the eye that reaches into your spirit. Every animal is not the same. Some do not want that bond and others actively seek

Shetty, my favourite bear, feeling a little shy while I was trying to do my piece to camera.
Photo credit: Craig Foster

it. I have spent twenty years around animals of all shapes and sizes and this is something I have seen in them. The first rule is that they have, just like we do, a sense of their own space. They have a comfort zone and this comfort zone is a space around them that they guard jealously. In the wild, if one respects this sense of personal space, it's the first step towards not being eaten or attacked. Land animals have a greater sense of this space than animals in the ocean. Animals in captivity have the same. Moving into their space and crowding them in that comfort zone is the quickest way to get attacked. Letting them get comfortable enough with you, such that they then voluntarily move closer to you, reducing their comfort zone, is the first step in getting them to trust you.

Shetty had been a fun playful young bear and he was in my space, practically on my lap, since day one. He had a habit of playfully gnawing on my arm and hugging it and as an adult bear, he does the same. That moment when they cross towards you, voluntarily giving up that comfort zone, is the moment of magic. It is also interesting to understand that just because they do it one day it does not mean

that they will do it again, if ever. Some of them always let you in once
that comfort barrier has been crossed and others never will, while still
others will do it depending on their mood that day. The day I did the
PTC, Shetty gave me the gift of letting me into his space and feeling
him hugging my arms and snuffling against my shoulder, I was so glad
he was not one of the damaged bears still standing on the side of a hot
road, dancing, and better yet, his old human family had moved on
with their lives as well.

Is it possible to rewild the younger bears? Perhaps. It takes a lot
of work and hope. There is a project in Northeast India, run by the
Wildlife Trust of India, another organisation dedicated to conservation.
They are attempting the rewilding of several black bears they rescued
from people who had kept them illegally as pets. Each of the bears has
one handler; they are in large enclosures in the heart of the forest and
are walked daily into the forest to acclimatise them to their new homes.
These walks and wanders also allow them to regain some of their wild
skills when it comes to recognising what they can and cannot eat in
the wild as once they are free, they will have to fend for themselves.
They are being released in protected areas, at least minimising their
contact with local people. Having been around people, they will not
have the same sense of self-preservation and fear that is essential for
any wild animal to survive. The bears have all been radio collared and
are still being monitored closely by their handlers. Only once they
become entirely free will one be able to gauge if the rewilding has been
a success. But even this semi-wild release is better than their living
as pets.

This does not, however, mean that it is possible for all bears to
be free. First, they will all need safe places where they can be released.
Secondly, they need to be released as far away from humans as possible,
and thirdly, many of them are too damaged physically to survive in
the wild. With over 800 dancing bears in captivity around India, it
is an impossible dream to expect them all to go home to the wild one
day. The closest they can come to it is places like this sanctuary where
they at least live out their days in a stress-free environment that is as
close to a wild forest as possible. The sloth bears introduced me to the
greatest lesson I learnt as a conservation journalist, the temptation to
make the 'other' the problem. It is easy to point fingers always at the

kalanders, the poachers, the local communities who live near forests as the problem. It is easy to bemoan how they are ruining all of it and how I am trying to help and save the wild. I could not have been more wrong. It was while doing the bear story that I understood the fundamental of a very essential lesson. That of the difference between livelihood and lifestyle. Livelihood is a matter of survival. Whether it's collection of forest produce, firewood, poaching of prey animals for meat, or cattle grazing, livelihood issues have solutions. Involve the people instead of alienating them, make the wildlife and wild spaces lucrative for them to make conservation worthwhile and nurture the already existing bedrock of love and respect for nature because in the 20 years that I have been a journalist, I can tell you that if anything has survived in this country it is because of the innate tolerance that people have shown. Lifestyle however is a far trickier beast. Lifestyle is how you and I live and the choices we make. We are not happy with any changes to our lifestyle at all, yet demand that people who have nothing change their livelihood. Lifestyle can be changed if you and I decide to do so.

The baby bears there were trapped because of a livelihood issue and their wild brethren are vanishing because of lifestyle choices. This is the lesson a dancing bear showed me.

When you are where wild bears live, you learn to pay attention to the rhythm of the land and yourself. Bears not only make the habitat rich, they enrich us just by being.
– Linda Jo Hunter, Lonesome for Bears:
A Woman's Journey in the Tracks of the Wilderness

5

ELEPHANTS
WISDOM OF THE WILD

The question is, are we happy to suppose that our grandchildren may never be able to see an elephant except in a picture book?
— David Attenborough

IN 2002, in West Bengal's Kurseong Tukra Basti, something tragic and then horrific happened. It became the scene of the only recorded man-eating elephant attack. The words man-eating and elephant are just too paradoxical to be in the same sentence, let alone be the moniker for one of India's most infamous rogue elephants. One of the world's largest herbivores a man-eater? Not possible, even in the imagination

of a B grade horror story writer. The female elephant killed over seventeen people before she was shot and her autopsy revealed that she had indeed consumed human flesh. Eyewitness accounts during the various attacks were dismissed as hysterical observations until the autopsy and DNA tests.

Man-elephant conflict in West Bengal is decades old. Lush forests have been destroyed for roads, canals, farms, irrigation projects and dams, vast swathes taken over by mining and growing human populations. Let's however not point fingers because all of the mentioned development projects bring us electricity, water, construction material and so on. You and I even have a direct link to this mess. It's in every cup of tea we drink. The scenic hills and valleys here, once home to lush forests and elephant herds, have over the years become tea plantations. The elephants have not only lost their homes, but even their migratory routes and have taken to raiding crops, bringing them into conflict with people.

This female elephant was part of a raiding herd and was being chased by people with torches, firecrackers, burning rubber tyres and even country made guns, practically every night. One night, it went horribly wrong. This young female was a mother with a nursing calf. Her hunger and fatigue were so great that she would often get separated from the herd and be isolated with her calf. On that fateful night in the chaos of the chase, her calf was shot and killed in front of her. Elephant mothers are dedicated, involved and deeply attached to their children much like we humans are. They have strong social bonds, intimate familial relationships and recognise friends and family across herds. Their young are the most precious members of the herd and are always protected and well cared for. For this young mother, the death of her calf combined with the daily stress of conflict and hunger flipped a switch that dragged her into an abyss of rage. She went o a rampage that killed seventeen people that only ended when a bull found its way into her brain.

Elephants can be brutal when they are in a state of rage. They v grab a person with their trunks, smash them into the ground, slam th against trees, throw them around, crush them by kneeling on them pressing down with their massive foreheads. Male elephants will gore the person with their tusks. What they will not do, thou bite. One can only wonder at the devolution in this female that she not only clearly bit her victims but actually consumed some flesh. The only explanation seems to be that the combined load of stress with the daily conflict and hunger and thirst had crested with the killing of her baby right in front of her, causing a literal psychotic break.

To me it is a huge screaming warning from nature about how unnatural we have made this living world around us.

Just a few years later, I would be sitting on the side of the road in Orissa, watching a baby elephant weep over the dead body of her mother. This female was one of the members of a raiding herd in Kheonjhar in Orissa and had been killed by the local people in revenge for the losses the people suffered with the destruction of their crops. This herd had also killed a few people in the daily conflict that has gripped this part of Orissa. The elephant was killed when she stepped on a cut off live wire, dragged down from a high tension cable and strategically placed on the path the herd normally walked. This is a particularly cruel and effective method that has killed dozens of elephants across India, and even some tigers.

The baby, hardly a few weeks old, was lying on top of her mother and nudging her with her head and trunk and making heartbreaking sounds of squeals, cries and groans. The forest department came to take the baby away and I sat down on the side of the road and wept. Four people had also been killed in and around that village that year by elephants and the people of the village were scared. No one looked happy or was gloating. Several villagers were also in tears. I was here to do a story on man-animal conflict and had certainly gotten more than I bargained for. In over a decade of reporting I had seen nothing like I would at Kheonjhar.

Growing up in Chennai, with regular visits to the local temples, elephants were a rather ubiquitous feature in my life. On weekends sometimes, in the local Guindy park that was a zoo, one could take an elephant ride. As a child, I was just super happy to be around these animals and did not think about the issues of captivity or how the elephants were treated. There was no overt sign of cruelty like wounds or illness and the handlers were not beating or hurting the animals in front of me. To a six-year-old, an elephant is a gigantic creature. It seemed like a being impervious to harm or struggle. The chains around

Elephants are very social creatures and show affection towards each other with various touches and gestures. Here two female African elephants are saying hello to each other.
Photo credit: Craig Foster

its feet, which it dragged effortlessly, seemed like nothing. They were like the silver anklets I sometimes like to wear.

It was uncle Siddharth who first alerted me to the fact that something was not right.

'It's such a shame,' he said.

'What is?' I asked, after I had narrated how Lakshmi the temple elephant had eaten bananas from my hand.

'She should be free, free to roam in a forest and be in her own world,' he said.

'But Lakshmi is happy,' I told him. 'She gets fed everyday and everyone is so nice to her.'

'Perhaps' he said, 'perhaps she is lucky and is treated well, but all creatures are born wild, Swati, and they should be free.'

On that day, it still did not make a great impression on me but in the years to come, those words would echo inside me, pushing me to do what I do.

There is an old story that tells of five blind men and an elephant. When the five blind men touched an elephant, one of them felt the trunk and claimed that the elephant was a snake, another on touching the legs said the elephant was a pillar, the third man touched the tail

In Kwai Conservancy, I had the opportunity to film several elephant herds and this was one group that would come very close to where my camp was.
Photo credit: Craig Foster

and said the elephant was a rope, the fourth touched the ears and said the elephant was a fan, and the fifth man touched the skin and said the elephant was a wall. All five of them were wrong but each of them went away convinced he was right. Perhaps it is a parable about perceptions but to my mind, in the literal case of the elephant, it is a clear example of the statement, 'There are none that are so blind as those who will not see.' What we as a people and as a co-species on the planet are not seeing is that this great mega species is in deep trouble. A world that seems to have no place for these sentient and ancient beings is a world that will soon have no place for us.

The summer of 2014, I stood in Satyamangalam Tiger Reserve in Tamil Nadu to do a report for the Save Our Tigers campaign on NDTV. Satyamangalam has just started a recovery after more than two decades under the destructive presence of the notorious Veerappan, a sandalwood and ivory smuggler. Veerappan started out as a thug, progressed to become a dacoit, and graduated to become a murdering, kidnapping, poaching scourge. The police force of three states – Karnataka, Kerala and Tamil Nadu – failed in bringing him to justice for years until he was finally tracked down and shot in 2004. By that time, he had murdered dozens of people, kidnapped dozens for ransom and killed over 500 elephants while smuggling hundreds of kilos of sandalwood, decimating thousands of old sandalwood trees.

Standing under the shade of a large banyan tree, in the cool wind of the early morning before the sun reached the highline, I could feel the spirit of elephants lost and gone. I could feel their large bodies in the brush of the wind, hear their breathing in the rustle of the banyan trees and in the dawn shadows before light chased the darkness, I could see their tuskless carcasses bleeding on the ground. Satyamangalam has a reputation for being one of the most haunted places in Tamil Nadu. Visitors speak of moving lanterns held by unseen hands, deep screams in the night and a general sense of foreboding. I know it's haunted. It's haunted by all the magnificent tuskers lost before their time.

In the Asian elephant species, only the male elephants have tusks. With hundreds of tuskers killed in their prime, Veerappan forever skewed the male female ratio of elephants in this part of India. Veerappan caused a counter evolutionary phenomenon of the makhna or the tuskless male elephants. With the prime males out of the equation, it was the weaker males left to mate with the females, and the result was a generation of tuskless male elephants.

Veerappan, however, as prolific as he was, was not the only elephant poacher. Elephants are poached across India and their tusks are sold for the insatiable demand for ivory in the far East. Asian elephant tusks are softer to carve than African elephant tusks and are therefore in great demand. This does not prevent large poaching syndicates from targeting African elephants, though.

My first story on elephants for NDTV in 1997 was about poaching for ivory and the effect it was having on Asian elephant numbers and the failure of Project Elephant to monitor and protect these animals. For several years after that, while poaching was still a problem, the greater issue was the loss of habitat. This one singular issue is the death knell. By 2014, the rate of habitat loss was even greater than in 1997; population numbers of humans and our greed and demand have grown. At times, the post for head of Project Elephant has been vacant for a year.

Today, there are roughly 20,000–25,000 wild Asiatic elephants in India. This is about 60 per cent of the world population of Asiatic elephants who are found in Sri Lanka, Bhutan, Thailand, Sumatra, Borneo, Nepal and Bangladesh. Classified as Elephas maximus, the Asiatic elephant is listed as endangered by the IUCN.

In India, they are mainly found in South India, in Kerala, Tamil Nadu and Karnataka. The North-eastern population is found at the Himalayan foothills of Bhutan and Northwest Bengal, eastwards into the states of Assam, Arunachal Pradesh, Nagaland, Manipur, Mizoram, Tripura and Meghalaya. They are also found in the Northwest in Uttar Pradesh and Uttarakhand and in central India in Orissa and Jharkhand.

Today, there are no wild elephants found naturally in Madhya Pradesh, but cave paintings from the famous Bhimbetka rock cave painting sites depict elephants.

In Orissa, the situation with elephants is so bad that they are moving into states like Chhattisgarh where they have not been sighted for generations. Some elephants have even been pushed into Andhra Pradesh from Orissa.

In the autumn of 2006, I went to Kheonjhar district in Orissa, which was once known for lush ancient saal forests with one of the largest elephant populations. The question I was asking was, how did a place that was once so abundant with natural gifts become a graveyard for elephants?

Nine elephants were killed in the month that we were there and over a 100 elephants had gone missing from the census counts in the last few years. While many were killed by poachers, many had lost their lives to the man-elephant conflict in the state.

For the local communities here, the elephants had become a nightmare. I spent two days in a village where 22 of the 23 homes were destroyed by elephants and the villagers slept in caves up in the hills, too terrified to sleep in their homes. I spent a night in the village. We climbed the hills before darkness fell and small fires were lit outside the caves for warmth. Meagre amounts of food were cooked, which they very kindly invited us to eat, and the children were already bundled into the back of the caves, wrapped in old torn blankets. There were no soft mattresses, no hot chocolate and marshmallows and snacks and other comforts we expect even when we are camping in the rough. This was not that. This was survival in its basic form. These people could not sleep in their homes. Grains stored after harvest are stored in the homes and the wild elephants were simply kicking their way in. Many of them would not even sleep here because they would be standing guard over their fields and outside the abandoned homes with fire torches and firecrackers to ward off the raiding elephants.

I met a man whose entire back was flayed. Every bit of skin had been scraped off because his run in with a big male elephant did not go well. He told me he was one of the lucky ones that night because his friend was killed by the same elephant. This man escaped by throwing himself on the ground and wrapping his arms firmly around the elephant's hind leg. The animals could not get to him in this position and only ended up dragging him all over, thereby scraping off all of his skin, before finally kicking him off.

An elder in the village told me that he had never seen anything like it. Growing up and for years after, the elephant was always worshipped. They were gentle giants who would walk like shadows through the forests causing no disturbances. He could not understand why his favourite animal was now so angry.

I had an answer for him. Hundreds of mining operations in the area had simply ruined the forests, destroying the traditional elephant corridors. Elephants live and move in herds. They are led by the eldest

Captive African elephants in Knysna Elephant Park. All of them are culling orphans and have been domesticated and trained to deal with tourists from behind a barrier. I had permission to interact with them freely.
Photo credit: Craig Foster

matriarch. She leads her herds through time-tested paths that they have used for decades. She will know where the best food and water resources are and even in times of drought, will lead them to water in a path memorised or handed down to her by the previous matriarch. When these traditional routes are destroyed, it throws the herd into disarray. The Asian elephant consumes close to 140 litres of water and 140 kilos of food per day per elephant. The devastated forests here, with the fly ash coating the leaves, clogging one's breath, large open mining ditches, water bodies thick and slick with tar are starving them out, forcing them to turn to other food sources, like the fields. The villages have also grown in number with a growing population and the cropping patterns have also changed, with fruit trees, vegetables and other crops replacing rice fields.

In the ten days I spent in Kheonjhar, raids by elephants on the fields were a nightly occurrence. Every night, we would stay up on the edge of the field and watch villagers light fire torches and patrol the fields. The elephants would come, looming out of the dark, silent and unstoppable, and the villagers would then beat drums, burst firecrackers and try to chase them away. I even saw elephant herds come out in broad daylight into the fields with their young, a very unusual thing for them to do as they are super protective of their young. Mobs of people would chase them and humans and elephants would run and scatter everywhere. It's dangerous enough in daylight, but at night, it was a recipe for disaster. The forest department seemed pretty helpless in the face of this onslaught and also had to face irate villagers.

I have never felt so angry and helpless on a shoot before. I was upset for the elephants and so sorry for what the people were facing. It also angered me that there was a general denial that the mining was the problem. I was given many explanations. I was told that the slash and burn agriculture pattern was causing this problem. I was told that the growing human population was encroaching into elephant territory. I was also told that the elephant population was growing and that they were leaving the forests. I am sure all of these reasons were part of the problem, but no one could answer one simple question – why was it that this conflict began in a big way only a decade ago? Up until then, people and elephants co-existed calmly with the occasional tusker in musth causing a problem.

Ten years ago, before my visit around 1995, the boom in Orissa's mining industry began. The price of steel and iron ore had almost doubled and the state decided to cash in. Hundreds of operations mushroomed overnight, many of them illegal with no environment clearances.

In an ideal world I would like to ban all mining, but that is not my world. It is also hugely hypocritical of me to ask for all mining to be banned in this world, as I, as a city dweller, live in an urban jungle filled with iron and steel. Mining unfortunately is a necessary evil unless we change our entire way of life. This we know is not going to happen and until then, our efforts need to go into ensuring that we can mitigate the impact on the environment. There is a lack of transparency in the whole process, starting from the way licenses are handed out, to corruption and a lack of social responsibility in the industry.

Here in Kheonjhar, it seemed mindboggling that there was no assessment done of the existing elephant corridors and measures were not taken to even acknowledge that there could be a problem with the elephants. While walking through the forest areas, several kilometres away from the main mines, I could see trees were covered in grime, the leaves were black, my face and hands were covered in fine soot, and the smalls streams and rivulets were muddy brown and black.

This area is also the catchment area of the grand Brahmini and Bhaitarini rivers, lifelines that flow through Orissa. The Brahmini and the Bhaitarini also power one of the greatest ecosystems here – the Bhiterkanika mangrove forests, the last mangroves left standing in Orissa. It is the second largest mangrove forest in India, after Sunderbans, and the last bastion of defence against the storms that build over the Bay of Bengal. Plans were on when I was there to divert more water from these rivers for bigger mining companies that had put in bids.

The greater shock was the vast tracts of old saal forests that were simply hacked down, not to clear land but to gain wood that would become the chassis of the thousands of trucks that are used to move the products from here to the Paradip port.

The government's own records showed the death of 300 people and over 300 elephants in just the last four years.

On being questioned, the mining minister dismissed all my questions by saying that it was not his problem and that the forest department was to blame for not sorting out the matter. He said he was in charge of ensuring development for the state. On questioning the forest department, I was faced with helpless hand wringing and lame excuses about how their recommendations were always dismissed and that the mining operations were unchecked and creating chaos. I even confronted the Chief Minister, Naveen Patnaik, who gave me the nicely practised reply of how he would look into the matter and that he would not allow the destruction of forests for development. In the four years of conflict, death, destruction and loss, not one politician had visited the area to even ascertain the truth on the ground.

Today, the new government is busy dismantling as many safeguards in our environment law as possible to push through greater development with mining at the forefront. It's a recipe for disaster that will only lead to more Kheonjhars.

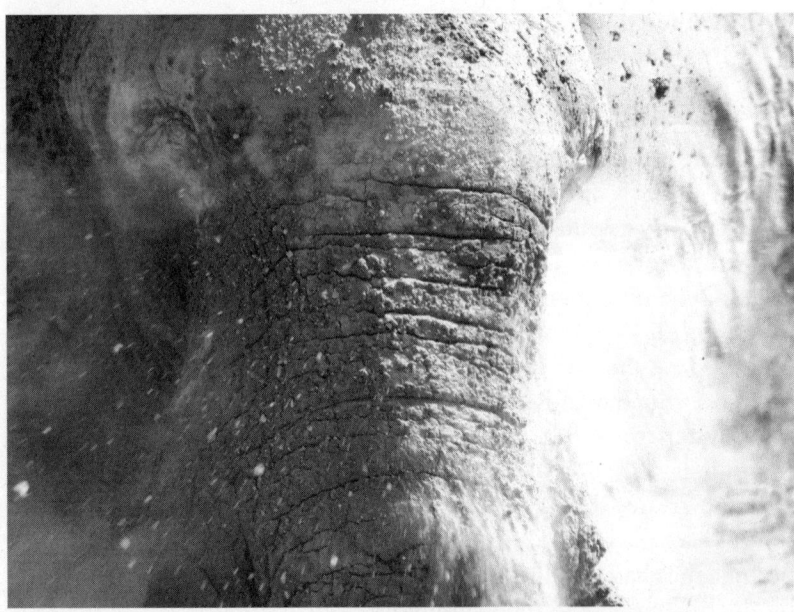

An old male African elephant dust bathing. They often dust bathe,
to both cool down and to get rid of small pests that make their skin itchy.
Photo credit: Tom Foster

Orissa probably has some of the largest numbers of mines in the country and thousands of crores are being made, but in the last decade the social indicators in the state are still abysmal. It remains one of the most underdeveloped states in India and a shining example of how these promises that sectors like mining will help in development are empty and bogus. In fact, states like Madhya Pradesh, Jharkhand, Bihar and Chhattisgarh and Maharashtra, which have some of the biggest mining projects, all have really poor social indicators. Things are only getting worse. 842 million Indians are malnourished in our country. Over 300 million live below the poverty line. I am not saying that some Indians have not gotten rich in this process but for those villagers sleeping in a cave and for those elephants starving and turning violent, it's pretty poor compensation. I have visited most of the places with big coal plants and I have seen the conditions of the towns. People can barely breathe, there is a noxious smog that hangs over everything. All the water bodies are either choked with rubble or soaked in tarry run offs. Temperatures soar to over 40 degrees in the summer with no shade or the cooling effect of trees. Ironically, some of these towns like Chandrapur in Maharshtra, one of the biggest coal plant towns, suffers up to eight hours of power cuts a day.

I find it hard to understand this model of development. What we have lost is incalculable because none of the things like the loss of forest cover, the loss of fresh water, the loss of biodiversity is accounted for in terms of money but crores of rupees are spent in health costs, crores in damage to crops, crores in floods and droughts and landslides and mudslides. For those of us in the city, we get our water from the turn of a tap, our electricity from the tap of a switch, our food from supermarkets, and we live our lives either oblivious to the horrific hidden costs or wilfully ignorant. One day, all the coal and other minerals will run out and that day we will stand on the graveyard of fallen forests without a soothing green cover or water knowing that as a species we are responsible for killing our own chance at survival on this planet.

Orissa is just one example in a rapidly deteriorating elephant landscape. In India, while most of the areas where the tiger is found is covered by national parks and tiger reserves, only 22 per cent of elephant habitat is protected. The elephant in India is a Schedule 1 species and is meant to be as protected as the tiger. Assam has lost close to 70 per cent of its forests since the 1970s, while altogether, the Northeastern states lost about 1,000 sq km of forests in the 1990s.

Elephants are distributed across only 3 per cent of India's geographical area. On an average, about 70–80 people are killed every year by elephants and hundreds of acres of crops destroyed. The central and state governments together spend about 10–15 crore rupees a year on controlling this depredation of crops and as compensation to people affected by the losses.

For the pachyderms, life has become excessively confusing. In conversation with Dr A.T. Johnsingh of the Wildlife Institute of India, one of the best-known elephant experts in the world, I learnt several things. The most important thing that stayed with me was Dr Johnsingh telling me how elephants are tormented when they kill a human. He said they feel deeply and are filled with confusion when forced into a confrontation. That made me sad. Elephants are one of the most sentient beings on earth. They love hard, have deep long-term social bonds, recognise family across herds, mourn their dead and will fight to the death for their young. They are nature's best gardeners, architects and guardians. Their paths cut natural fire lines through forests, their feeding reduces forests to grasslands that become home to hundreds of other grazers, while their dung spreads across forests and feeds entire ecosystems of insects, butterflies and birds. Their wisdom is ancient and moves out of synchronicity with the way we view time.

To an elephant, an overgrazed forest today is a grassland for a generation in the future. A grassland covered in their dung is a forest in the future. We don't understand their reasons nor do we appreciate their rhythms. The elephant lives outside of the human conception of time. To us, everything needs to fit into the neatly divided hours, days, months we have segregated into measuring our lives. For the elephant, time is measured in the generations to come, to the web of life she is so wholly a part of. This basic divide between us in understanding time is always going to make us react to them as if they are forces that need to be controlled.

We build roads, dams, homes, fields, mines, bridges, railway tracks, just cutting a swathe across the landscape and then are surprised when the elephants come kicking down our doors.

Every year, with great pomp and splendour, we deck out our Ganeshas and go on what I term a rampage of religious drama, ruining streets and rivers with garbage and toxic chemicals, in a way perhaps that reflects how we treat our living Ganeshas. In Orissa, on my last day, I watched as a group of villagers covered the dead mother elephant in a saffron cloth, laid marigold flowers on her body, painted a large tikka on her forehead and then buried her. There was some prayer, some chanting, but all I could remember was her crying baby, born wild, but destined for a zoo now. It felt grotesque to me like some horrible pageant gone wrong. Years later, I would be watching another dead elephant under entirely different circumstances.

In 2007, after I first came to Africa, I went on a long journey with my filmmaker husband, Craig Foster. He was making a film on the Khomani San, the last of the living hunter gatherers of South Africa. The journey brought me to Namibia and there I was, sitting on a tree, watching a dead elephant.

We were in a small conservancy in Namibia, in Tsumkwe. This area has several herds of wild elephants and allows for the trophy hunting of a few every few years. The hunting is very regulated with hunters being only allowed to pick an old or sick and infirm elephant. The elephant had to be well past its reproductive age, had to have left the herd or been abandoned and down to its last molars. Old elephants usually have very worn-down teeth after a life time of chewing. When they are down to their last few molars, it becomes difficult for them to eat as much as they need to stay alive and they become weak and die from starvation. It's not unpleasant or painful, just a slow shutting down of organs and life. It is sometimes ironic that the very idea of hunting will cause a massive hue and cry in the general population, while the idea that tourism can be quite ecologicaly damaging will not even be entertained. In reserves where elephants are not allowed to be hunted, the main money is made from tourists. To house tourists, lodges need to be built, roads need to be made, vehicles need to be driven. Tourists want hot water, fresh food, swimming pools, air conditioning in the summer and heaters in the winter. All of this adds up to an enormous footprint on the wild ecosystem. We have seen this in most of our national parks in India with very few tourist operations being truly green and sensitive to the environmental concerns of where they are situated. In Namibia in this conservancy, hunters have no lodges. They can either organize home stays, which means they stay with a local family or camp. Usually they camp as they have to track the elephant they are assigned to hunt. This could take a few days or more. There are no amenities available in the jungle while they are tracking the animal. Once the elephant is spotted it is shot.

I am not advocating trophy hunting of elephants, but what I saw in Namibia seemed to be working. It just occurred to me that not all conservation answers were black and white. I also realized that every area, every population, every culture had its own successes and failures and the attitude that there is a one size that fits all is the worst for conservation. Hunters pay thousands of dollars to shoot the elephant and the money goes back to the villagers who have a stake in the conservancy. Ironically, with the creation of the conservancy and the regulated hunting, poaching here is non existent. The villagers know their great profits lie in only allowing the elephants to grow old and tourists come to see the other elephants.

This elephant had been shot through the temple, a clean shot that had dropped it almost immediately. While I sat on the tree and watched, hundreds of villagers arrived at the scene. Unlike in India, there was no ceremony. They all just pulled out knives, some of them smaller than my butter knife, and started to carve up the elephant. I nearly fell out of the tree. I asked Craig if this was normal and he said, 'Well, it's dead now, so they will honour its life by eating it and it does not go to waste.' Within a few hours, there were just bones left and people had walked away with the meat. Several families would live on this meat for days, if not weeks. Craig and his brother brought some of the meat back to camp and cooked it in a stew. I just could not bring myself to eat it, not because I thought it wrong, but because I guess all my life I have had a connection with Ganesha and just felt downright weird at the idea of eating elephant meat. But I could well see why everyone else did and felt ok that it was so. This elephant had lived out his entire life in the wild as he should, died very quickly, and helped hundreds of families who otherwise cannot afford too much food. It felt like a greater tribute to an animal than the tribute given to the young mother who died by electrocution in a revenge killing, left her calf orphaned and was then just put into a hole in the ground. I found no sanctity in that.

An African elephant herd drinking water from the Savuti channel right in front of my tent.
Photo credit: Craig Foster

The African elephant is much larger than the Asian elephant. Both males and females have tusks and they have much larger ears. African elephants also have a reputation for being fierce. In Africa, the elephants face the same problems they do in India, from habitat loss to poaching. The poaching, though, is on a far larger scale. About 30,000 are poached per year. In just four years, close to 250,000 of them were killed by poachers. It is estimated that there are about 400,000 elephants left in Africa, so, if this is not stopped, the African elephant will be history within a decade.

Ivory. For this one substance, so much slaughter.

In Africa, they also employ methods of elephant population control that are not practiced in India. One of the most controversial methods of elephant population control is culling. Culling involves the killing and removal of a certain sustainable number of animals from a given population to keep their numbers in strict control. In South Africa, all the national parks and protected areas are fenced in. This has caused problems with growing animal numbers in these areas out eating the food sources. Elephants being big feeders are especially vulnerable to being fenced in. In the natural course of their lives, they will move over vast distances to feed so they do not decimate any one area too much. In South Africa, because of the fences, a growing population has nowhere to go and end up overgrazing areas, putting other herbivores in jeopardy by removing their food sources. Instead of removing fences and encouraging natural movement, the authorities instead use culling as a method of controlling numbers. In a dry scientific discussion, this might sound like an ok plan, but for the elephants, it's a nightmare. Because of their deep social and familial bonds, culling is a massive trauma for the herd. In the old days before better understanding of elephant behaviour, random elephants would be picked and killed. No one understood that this would destabilise herds and cause havoc. Now, they try to eliminate entire herds. Culling was stopped in South Africa in 1995 due to a growing public outrage with the practice, but with a growing elephant population trapped in smaller and smaller areas, culling has resumed in the past few years.

What is worse is that many of the younger elephants are not killed with their elders and are sold into domestic captivity or zoos. It might sound like a kind thing because at least the animals were spared death, but in reality, it is a very problematic practice. The orphans are traumatised by witnessing the deaths of parents, aunts and uncles.

Elephants drinking water. Their trunks are both unbelievably strong and incredibly flexible.
Photo credit: Swati Thiyagarajan

They often carry this trauma all their lives just like children who survive wars do. In the elephant social structure, the younger elephants are taught and controlled by the older elephants and the elders. Just like in a human family, when the older brother or sister or even an aunt and uncle might discipline and teach a child how to behave. They are lost without that love and guidance.

Culling also makes other elephants who were not part of the cull, angry and aggressive. They are very aware of everything that is happening and can communicate with each other over very long distances through both acoustic communication (that is the various sounds they make) and seismic communication. In seismic communication, they make a low rumbling sound that is not just heard but actually vibrates and travels through the ground. Other elephants can pick up this sound both with their big ears and also through their very sensitive feet and toes and the tips of their trunk. Usually, during a cull, teams try and not cull herds that are less than five miles away from other elephants.

The other practice employed to control elephant numbers is contraception. This is not without controversies. Firstly, amongst elephants, it's not just the females who come into heat. A male elephant in heat is said to be in musth. This is when the testosterone levels in his body go through the roof. He has so much testosterone in his system that it constantly dribbles out of him with his urine through the day. An oily secretion is also seen pouring out of the glands on his face. Elephants in musth are aggressive, unpredictable and best avoided. So if cows, that is the female elephants, are on contraceptives, then they will not be in heat and therefore unavailable for mating with a bull in musth. This makes him even more aggressive and dangerous as he has no outlet for what nature intends for him to do at this time. Some male elephants can stay in musth for up to nine months. It also has to be administered every year if it is to have an effect in sexually mature females, and it has to be stopped for a few years to allow for the birth of fresh calves so as not to disturb a herd's balance too much.

It is only now that a contraceptive is available for the male bulls as well. The delivery mechanism is to shoot the animal with a dart and this has its own problems. The animals become wary, angry and confused when they are chased by a helicopter with shooters that dart them. Contraceptives are also expensive, not to mention all the preparation and manpower that is required to administer it. Recently the West Bengal government mooted plans to use contraceptives in certain elephant populations in the state due to the rising conflict between people and elephants. The Supreme Court had to intervene to call the idea crazy. However, the forest department in Karnataka have just decided to use birth control on the elephant populations in Karnataka.

Sometimes, one other method is employed – translocation. This, however, is riddled with issues. Firstly, who chooses which elephant or group of elephants gets translocated, even if space is available for their translocation? Secondly, it costs close to 25,000 dollars to translocate a single elephant. Today in Africa, there is no space for these animals outside of existing protected areas and most of those areas have shrunk and are already dealing with more elephants than they can handle. Small private game reserves do accept a few elephants every year, but not more than two or three, as usually they may not have the space for more than ten elephants and need to allow the translocated ones to settle in and form their own herd by having babies. This is not always successful as without older matriarchs and a middle layer of aunts, young females do not learn to be good mothers. In an elephant herd, young females often help babysit the calves and look after them under the supervision of their older sisters, aunts, grandmother and other females. Years of doing this and observing those females gives her an idea of what to do when she is ready to have her own calf. Often, when there are problems with elephants in captivity or in private game reserves, it is because these elephants have suffered a trauma that required the translocation in the first place and also have had not seasoning and tempering or teachings from their older wiser herd mates. With male elephants, the older members who have left their family groups and wander about on their own become the nucleus of a bachelor group. Young male elephants ejected by the matriarch from a family group, join such males. The older male teaches the youngsters how to behave and be 'grown up' elephants. The young males then mature and rejoin females to mate. They will continue to visit with the older male and sometimes stay until the older male dies and then go onto becoming the wise old men in their turn training other young male elephants. These young males are called Askaris, in Africa.

Almost forty years ago, a new man-made park was created in the Pilansberg. The Pilansberg, an area of small valleys and hills was a space created by mining. Ninety miles from Johannesburg, the area was cleared of all non indigenous vegetation and people and made into an animal reserve. In one of the biggest wildlife translocations, over nineteen species of animals were introduced into the park in an attempt to naturalise it. Amongst the introduced animals were young elephants. All of them orphans from various culls over the decades in the Kruger National Park. For the first few decades, it all seemed to go ok as the elephants were young and settling in. As these elephants grew older, problems started to set in.

In the 1990s, these young adult male elephants not only started to charge and attack tourist vehicles in the park, they also started to kill the rhinos. These young males seemed to have come into sexual maturity much earlier than they should have and as the older females would not accept them, they had a lot of leftover aggression and attempted to mate with the rhinos. It was only after a few adult bull elephants were introduced into the park that their behaviour sorted itself out. The young adult males seemed to have come into musth early as there were no adult males around. They were also in musth for a lot longer than is normal, increasing their levels of aggression. The cull experience had also left them very wary and angry with people. Once the adult male elephants arrived, the musth cycles slowed down and there was an automatic calming in the young bulls. Since the late 1990s, there have been no rhino killings or tourist attacks in the Pilansberg.

This proved beyond a doubt that a herd is not just a group of elephants. It is a structure of elephants of all age groups working in tandem. This is what makes domesticating these animals and introducing them into new family groups difficult. The combination has to be just right and the animals need to get along. It's these intertwined connections, deep bonds and elephantine culture that makes it so difficult for us to even attempt to manage them.

And our growing intolerance is leading to growing conflict. An endless cycle of conflict that neither side is winning. In 2014, a rampaging bull elephant again in West Bengal went into a small village and kicked his way into a home from the back. The owners, a husband and wife, were eating dinner in the front room. Hearing the commotion, they panicked because their baby was in the backroom that was being kicked in. They watched in horror as the walls crumbled and fell. The elephant, however, suddenly stopped kicking in the house when he heard a sound. It was the baby crying. He looked around and could not see it. He then systematically removed every piece of rubble that had fallen on the spot where the baby was and until he fully uncovered the baby by removing the last piece of plaster, he did not stop. Once he made sure the baby was ok, he walked away. The parents, in awe of what they witnessed, only said 'It was like seeing Ganesha.'

There is also a relatively new phenomenon in Africa and that is the issue of captive elephants. As a tradition, the capture and training of elephants is an old one. Elephants have been captured and used since 4000 BC and the Indus valley civilisation. Elephants were used in battle, used to carry the royal entourage on hunts, and later used in the timber industry. Many of them were pressed into service in temples and now even though the machines have taken over the jobs elephants did for the timber industry, they are still used in temples, religious festivals, weddings, to ferry tourists and in safari parks in India. In Africa, the African elephant was considered too wild to be used like this and it was never a tradition. In Europe, however, both Alexander the Great and Hannibal used elephants in their armies. However, now there are many elephant tourism outfits where elephants orphaned or injured by culls are taken in. Many of these places allow for elephant interaction with people.

There are thousands of elephants in captivity across Asia, and in India, just Kerala alone has close to 700 elephants in captivity. By the early 1980s, it had become illegal to capture elephants in the wild in order to domesticate them. In India, the tradition of capturing elephants in the wild is called 'mela shikar'. In north India, female elephants are the main targets for captivity as they never go into musth and this was essential for the elephant armies. In south India, male and female elephants are used.

In 2003, a horrific incident captured the world. World famous elephant trainer Parbati Barua was caught on camera, attempting to train a young elephant. This elephant had been captured along with others from an area of Assam where his herd had been in conflict with people. The elephant was filmed being tied down and beaten with

African elephant trunk.
Photo credit: Craig Foster

bamboo sticks and prodded, and he later died. Parbati Barua, who
grew up with elephants, was called the 'queen of elephants' by Mark
Shand. Parbati who is an elephant expert, and someone who has trained
dozens of elephants suddenly found herself in the limelight for all the
wrong reasons. While there was no question that the film clip was
shocking, it was also a representation of one extreme form of trying
to 'break the spirit' of what was probably quite an aggressive young
male elephant. Unless the elephant feels a certain fear and respect for

his or her trainer and succumbs to their control or command, they are dangerous to have around. But this logic in itself is problematic and needs to be questioned. Parbati Barua is not someone who hates elephants; in fact, she is someone with great love for them, having lived most of her life with them. She just sees it as something she has become an expert in doing from childhood. She believes that wild elephants should be protected and believes that their habitats need saving. She also equally believes in the process of taming elephants required in captivity. It is such an old culture in India practiced by generations that no thought is given to the mental well being of the elephants in question. Again not out of inherent cruelty but almost like its seen as a necessity for a certain period until the animal is docile and open to orders.

In Africa, now there are places where African elephants have been trained enough to carry people on their backs and interact with the tourists.

In order to 'gentle' an elephant into being accepting of people in close quarters and of their handlers when given instructions, the regime is quite brutal. A lot of it involves 'breaking of their spirit' in order for them to accept the human as master. Often, on elephant back in forests in India, I have watched as mahouts use an iron-hooked implement to beat the elephants about the head in order to get them to obey instructions. It is uncomfortable to watch and I always cringe. In various Indian tiger reserves, the elephant is used to patrol the forests to get to places that vehicles cannot approach, so it seems like their service is invaluable and it is. The question here is – is it fair to them to have to be in this position in the first place?

For a long time, elephants in India were captured in the wild and the process involved using other domesticated elephants to herd the chosen wild elephant and corral them into a fenced in kraal. Here, the elephants are kept confined for days with a regime of 'harsh love'. A lot of beating, isolation and food as reward for good behaviour. The animals are captured young and so they have a better chance at being trained.

In many instances, the handlers and mahouts truly love their charges and have extraordinarily intimate relationships with them. There is a trust and affection between both man and animal. The problem lies in the fact that they are big animals and even the slightest loss of temper can turn dangerous. Just a swat from the trunk can mean death, if not several broken bones. This keeps the chain of 'breaking their spirit' in place in the hope that the more cowed they are by their handlers, the easier and more malleable they will be. Many of the elephants in captivity suffer from immense stress. It is a form of post traumatic stress syndrome, seen in humans who have faced horrific incidents like murder, war or other life threatening moments. It takes a toll on the animal and even an elephant that seems docile or friendly for the longest time can suddenly crack under pressure. During temple festivals and performances, the immense crowds, the noise, the confinement and also the extra rigorous training that takes place for weeks prior, can all contribute to this breaking point. They are chained for hours in a single position. They have to walk on cement pathways and paved roads that are really hot in the harsh summer, ruining their sensitive feet. Often elephants in these celebrations are animals that have been captured. They are animals that have been ripped away from family members and even animals that have been abused. Nearly every year in India we read about an elephant 'running amok' and killing

A baby elephant playing with his mother and aunt in the Kwai river.
Elephants love water and are excellent swimmers.
Photo credit: Craig Foster

Old African elephant male dust bathing.
Photo credit: Tom Foster

people in the crowd. This is the animal crying out desperately for help. There are more than 10,000 festivals, parades and other celebrations that take place in Kerala every year and the elephants are rented out for them. Over 200 people have been killed by captive elephants in Kerala and animal activists allege that over 1,000 elephants have met their deaths in the same amount of time.

In South Africa, I met a beautiful gentle giant called Namib, a rescue from a cull who killed his trainer. Post that he went back to being gentle for years and I even spent a whole day with him when I was filming him for a story. He was so friendly, gently nudging me with his trunk, allowing me to sit in his shadow when the sun got high and feeling in my pockets for treats. However not for a moment did I forget he had killed a person and a few months later he attacked and badly injured another trainer. After that he had to be moved from the park meant for tourism and sent away to a semi wild area where the elephants have no interactions with people and get to just be.

A few years ago, a group of elephants were in violent conflict with people in Hassan in Karnataka. Several people were killed by the elephants while several elephants were poisoned by irate people. Hassan was not an area used to elephants. Initially when the first few elephants turned up in the 1980s, it was a novelty factor. Soon as the numbers increased and the hungry animals started to destroy fields people started to get nervous and then angry. Then people were killed. For years as it was building, it seemed like local authorities were indifferent to the plight of the people. All the people seemed to want here was a sympathetic ear and some compassion for what they were dealing with. When that did not materialize and the conflict turned violent, the people too had had enough. The matter was brought to the attention of the courts and in 2014 the courts declared that the elephants had to be captured and removed from the area. Elephants like most other animals are territorial and have established ways of moving through their territories. They cannot just be chased from one area into another. Being incredible large animals that require space, food and water, they can't just be released into just any forest area. The capture was traumatic for everyone concerned. The forest department had no choice but to follow the orders of the court and the people could not be blamed for wanting the animals to go. The courts thought they were saving elephant lives as they were being killed in the conflict. So it was a bad solution to a bad situation or rather a situation with no good solutions. Already trained and captive elephants were used in the capture of these wild elephants. It took days and tears and stress. The wild elephants were suitably traumatised and the captive elephants were both belligerent with them under the orders from their mahouts and gentle when they could comfort their wild brethren. Just seeing the video footage of the capture filmed by a friend was enough to make me cry, but his footage also showed me how much the people had suffered and how much they had put up with and for how long they had been ignored in their plight. These elephants were then moved into kraals and their 'gentling' began. For the elephants there will be no happy ending and that is a fact. Horrific hidden camera footage of the elephants used in Guruvayur temple in Kerala has shown immense abuse and cruelty in how the animals are maintained and treated. I have been to many places where the elephants are well looked after, are allowed to roam free for long periods of time, albeit with chains around their legs, like in our various national parks, where without the elephants much of the protection and patrolling of the parks would be absolutely impossible. But for the few well-treated pachyderms, the numbers who are stressed, miserable and badly treated are legion. As a civilised society in the twenty-first century, it is appalling that we still have these beautiful animals or any animals at all in captivity.

There are several orphanages and sanctuaries now that cater to abused and rescued elephants from zoos and circuses and cruel owners who try and give them as much of a free life as possible. This is crucial as many of these animals really have nowhere to go.

I will never forget a female elephant in the Rajaji National Park. She had been taken for her walk as usual by her mahout, one morning, and then was asked to assist in the rescue of a heavy load vehicle trapped in soft mud. Ropes were attached to her and the vehicle and she was commanded to pull the vehicle out. In the soft mud, unable to get enough traction and with such a heavy load pulling her down, she broke her legs. She somehow managed to walk back to camp with her mahout before collapsing, which was when they even noticed that there was a problem. Once she collapsed, it was impossible to get her back up on her feet. An elephant that sits down for too long starts dying a slow death because her immense weight starts to steadily crush her own internal organs. An elephant with broken leg bones is almost beyond saving as without all her legs there is no way she can support her immense weight.

The debate then started on whether to euthanise her or not. Top vets said that she could not be saved and that euthanisation would be best and most humane. Animal rights activists agitated against the decision and said that more options including artificial leg braces should be explored before euthanising her. I spent a day and a night with this elephant who was just lying down and I could see her terrible pain and suffering in her eyes. I could not believe that there was even a debate. At one point, while I was stroking her ears, she leaned her head into me and I wished I had the lethal injection on me to administer it to her. It also made me wonder how many elephants like her were facing day-to-day suffering and had no voices to speak for them. Finally she was euthanised, but only after four days of horrific suffering. It was on that day that it became crystal clear for me. The practice of keeping captive elephants had to be banned. We simply cannot justify it.

In Africa, too, many animal rights organisations are fighting against elephants in captivity. As in all things, big money is involved and banning it might be next to impossible.

Elephants are a mirror to the best part of who we can be, wise, gentle, yet tough. Guardians of a world that is to be tended and passed onwards. However, in reality, our conflict with them is a reflection of who we have become today, that they face cruelty, deprivation, loss of home and indeed, are facing the path of their extinction.

But perhaps the most important lesson I learned is that there are no walls between humans and the elephants except those that we put up ourselves, and that until we allow not only elephants, but all living creatures their place in the sun, we can never be whole ourselves.

– Lawrence Anthony, *The Elephant Whisperer*

6

PELICANS, PAINTED STORKS AND PENGUINS
THE WILD THAT LIVES WITH US

Honor the sacred. Honor the Earth, our Mother. Honor the Elders. Honor all with whom we share the Earth: four-leggeds, two-leggeds, winged ones, swimmers, crawlers, plant and rock people. Walk in balance and beauty.
— Native American Elder

IN 2005, on a crisp winter morning, with little wisps of mist brushing the road as the sun rose, I found myself on a road from Bangalore to

a village called Kokkare Bellur. The word 'kokkare' means storks. The village I was heading to was known locally as the 'village of storks', as for over a hundred years, this village has played home to painted storks and pelicans who come here over the winter to give birth to their young. I thought it sounded wonderful and wanted to see for myself how much of it was fact and how much fiction.

As sunlight started to flood the world around me, I could make out that I was in the heart of rural Karnataka, where green fields, graceful trees and ponds streamed by outside my car window. Colourful children waved as we drove by and women went about their work busily. We should have been getting close to the village from the directions we were given, but I could not see any birds either in the trees or in the sky. I wondered if we were a bit lost or if we were further from the village than I had originally estimated. We took a bend in the road and high above me I saw the distinct shapes of pelicans. A huge sign said Kokkare Bellur, just as I started to hear the call of painted storks. Suddenly, every tree was filled with birds and the cacophony was both music and amazing. I had literally driven past hundreds of empty trees, but from this village border onwards, every tree was full.

The elders in the village would later tell me that the birds had been visiting them for over a hundred years and that even when the village moved location after an outbreak of the plague decades ago, the birds followed. It might sound astounding but for the fact that I had driven by dozens of villages without seeing a single bird, and yet, here there were hundreds.

To me, this was an important story. Not just because of the birds but because of an obvious link between the birds and the local people. To me, Kokkare Bellur was one of the last symbols of all that is still magical about India, the great tolerance and reverence for the natural world around us. The mega diversity this country holds despite its growing galloping problems only stands on the foundation of the tolerance and reverence of generations past, which is eroding fast in our race to 'develop'. Everywhere I have travelled I have seen that the forced separation between the people and the natural world around them has led to conservation setbacks, while integrating the local people has lead to conservation successes. Here in this village, it was not the government, not the forest department, not even policy that kept these nesting birds safe, but just the local people and their love for the birds.

The older generations that started this tradition saw the birds as the daughters of the village, girls who came home to give birth. They saw the birds as symbols of fortune and good will. A few years ago when the birds did not come, the rains did not come either and it was a bad year for the crops. It made sense as both the storks and the pelicans are birds that come to nest after the monsoons. For six months of every year, the village puts up with really loud and deafening bird cries, bird poop falling out of the sky, falling feathers and general confusion. They are quite indulgent of this. Especially the elders, who smile at me and say, 'They are like children you know, naughty but mean well.'

The birds, the painted storks and the spot-billed pelicans seem to be everywhere, their large nests weighing down the trees in the village.

The painted storks are large wading birds found in the wetlands of India and in some parts of south east Asia. It is their distinctive pink tertial feathers that give them the name of painted storks as they look like brush strokes on their bodies. They forage in flocks along rivers, ponds and lakes and their actions of immersing their half-open beaks in water while sweeping their heads from side to side not only allows them to snap up fish they feel by touch but also stir up the sediment, allowing other smaller birds to feed. They also use their feet to stir up the water, scaring the fish out of hiding. They practise colonial nesting and often nest along with other water birds. Here in Kokkare Bellur, they nest with the spot-billed pelican. Painted storks are not particularly migratory, moving from place to place depending on weather and food. The only time they are truly stationary is during nesting time. Juvenile birds leave the nest and travel quite far. One juvenile was tagged and recorded as having flown 800 kilometres away. Male and female painted storks are similar to look at with the males being larger than the females. While they don't necessarily mate for life, pairs may stay together for several mating seasons.

One of the beautiful secrets of the animal kingdom is the phenomenon of sexual monomorphism. This is when the males and females of a species look identical to each other without immediately distinguishable differences. Of course, there will be sexual differences marking them as male and female but no obvious immediately discernable difference to the eye. Usually, such animals pair up and stay monogamous. Male painted storks are slightly larger than the females but otherwise they are quite identical looking. Sexual dimorphism is when there is immediate difference between the males and females, like with the peacock. Animals that have similar looking males and females often pair up and stay monogamous for several seasons, if not for life. The others, like with the peacock, have multiple partners. Often, the mated pairs use the same nest built over several seasons as well.

Baby chicks are entirely dependent on their parents for food and eat the regurgitated fish when the parents fly back to them. Each chick requires about 500 grams of food a day. They are loud and demanding and grow quite big even before they can fly. When threatened, they will vomit out their food and crumple to the floor of the nest, pretending to be dead. Really young chicks are predated on by eagles. Before their feathers fully develop, they are very vulnerable to heat and cold and require one parent to be around them. In the heat of the mid-day sun in Kokkare Bellur, I watched several parents stand up spreading their wings over the nest, so that they could shelter their chicks.

The largest secure population of these birds are found in India. Birds in Pakistan along the Indus river system are endangered and chicks at their nests are taken away for the bird trade. The species was nearly decimated in Thailand while small populations are known to exist in Cambodia and Vietnam. With disappearing wetlands and climate change messing with rainfall patterns, painted storks have a lot of challenges to face today in India as well. They are classified as a near threatened species, making every single nesting population and site crucial to their continued health, and the tolerance and acceptance they find in Kokkare Bellur is very important to their future. The spot-billed pelicans are classified as vulnerable or more threatened than the painted storks. Habitat loss and human disturbance are the main reasons for their reducing numbers.

Like most other pelicans, they catch the fish in their huge bill pouch and just like the storks, they carry the fish back to the nest to feed their young. The young can stay in or near the nest from three to five months and are again very demanding, and the parents spend several hours a day flying back and forth with food, also having to feed themselves. Nesting colonies are found in India, Sri Lanka and

Cambodia with the maximum numbers of the birds found in India. They used to also be found in the Philippines and Burma but have since become locally extinct in several sites due to habitat loss. Again, the loss of a single crucial nesting site can quickly decimate their numbers.

While Kokkare Bellur is so far holding strong, things are changing. For one, the trees the birds use for nesting are valuable for their timber. For the birds, it is very important that the trees they nest in stay intact as many of them will use the same nests from previous seasons to nest in. If the trees get cut down, the nests get lost. Also, with fewer trees, there tends to be overcrowding in the existing trees and that can also cause problems. Growing numbers of people in the village means needs are also growing. The main occupation of the village is breeding and growing and harvesting silk worms. To feed the worms, leaves are required. The trees need their leaves to stay intact because without the leaves, there are two problems. One, the nests are left unprotected from shade and this can kill the chicks, and two, chicks also fall out of the nests without the leaves as a barrier. Wood is used for firewood, and also used to fire kilns in which bricks are baked for construction. In the old days, the houses were made of mud and had thatched roofs; now, the houses are made of brick and cement and require timber.

A young African penguin. They are called 'Blues'.
This is before they develop their black and white waterproof feathers.
Photo credit: Craig Foster

Realising all these changes, K Manu, a member/founder of the Mysore Amateur Naturalists Society, who helped me find this village and do the story, has dedicated his time to working with the villagers to keep the culture of conservation alive here.

It's not as if the birds only take and give nothing in return. For one, their presence guarantees a good monsoon as instinctively they will not breed during a drought period, so they are like an early warning system of weather patterns. Their droppings, known as guano, are a very rich natural fertiliser that is used in the fields for the crops. It is a better option than pesticides and chemical fertlisers. Standing under one tree, I also had a fish fall on my head. Manu told me that that was quite common as the parents bringing fish back to the chicks often dropped a few. These fish are fresh from the lake and the villagers usually collect these fallen fish and fish parts for food. Manu is trying to make them realise all these benefits so that the people keep continuing to support the birds. In the monsoon rains and squalls, often, baby birds fall out of their nests and the children of the village run around collecting them.

Manu has a small rescue centre where he looks after the chicks that are injured and cannot be returned to the nests. He nurtures them until they are ready to fly on their own. All of this has been done with his own money and some private grants. The government has contributed a little by compensating villagers for every tree they do not cut down. However, long term viability of the people's good will and the conservation of these beautiful birds rests on the area becoming tourist-friendly. Demands for small eco tourism huts, centres and cafes for food had not been met and some of the villagers told me that it was frustrating to not be able to attract a larger number of people to the village than just a few interested birders. Tourists would bring much needed income and thereby create greater incentive for the protection of the birds.

Tradition and culture, while crucially important in creating the bedrock for conservation, cannot feed hungry people. A growing India demands more. The bright city lights are coming closer to our rural areas and taking over. A changing India therefore needs changing solutions to conservation challenges. This village is just one of the five nesting sites of the spot-billed pelican in the world. The success rate of chick survival even with the availability of food water and shelter is just 1.6 chicks per nest. Places like Kokkare Bellur need to be preserved if spot-billed pelicans and painted storks are to survive in modern India.

The older generation had a love of the birds that the younger generation has to cultivate in between going to school, travelling, finding jobs outside the village and the advent of consumerism. Standing beneath a tree heavy with birds, some of the chicks nearly as big as their parents staring down at me, I could not help but wonder as to whether a decade from that day in 2005, if I came back for a visit, I would be greeted by the same sight.

I watched the adult birds soar on thermals high above in the sky, just dots of black against a blue so clear it hurt my eyes, and I wondered at the magic of this place, this village. These are birds that require a lot of fish both for themselves and their chicks, which requires hours of foraging in ponds and flying back and forth. There are no ponds close to this village. The closest pond is quite a distance away. What then brought the birds to this village? This spot? Why not other villages on this route which also had trees? If what the elders told me was true, the birds had even followed this village when they moved from their original spot to this one. That would have meant building new nests and re-establishing colonies.

What made these birds come here year after year? What makes them choose this village? These trees? I did not have answers on that day but in the years since, I have come to see amazing things about animals and people and I now know the answer. Animals can tell when they feel safe. They know when they can trust a person or a place. These birds are intelligent enough to predict the rains more accurately than the MET department. These birds come back to their nesting sites where they were born with unerring instinct. When their innate instinct and intelligence is that sharp, of course they know when they can trust a place and a people. They know when they can trust a space, trust that it will stay safe for them to come back.

Years from that day, I would stand halfway around the world on a windswept beach in Cape Town and watch another bird and marvel at both the similarities and differences in their conservation.

On my first day in Cape Town, sitting at a window in an idyllic spot in a house on a beach, a certain sight nearly made me fall out of said window. Waddling down the lawn towards a rocky outcropping in the ocean were four penguins. Here, I have to add some facts that might be known to all except Walt Disney. Penguins are strictly southern hemisphere birds. They do not occur in the Arctic. Or anywhere else in the northern hemisphere except in zoos. So any picture of a penguin and a polar bear standing cheek to cheek is photoshopped. In the southern hemisphere, they occur in Australia, South America, South Africa and the Antartic. There are seventeen different species of penguins in the world.

The birds outside of my window were African penguins, one of the more endangered penguins. Their waddle to the water was not very graceful and I wondered how they would negotiate the rocks with just those webbed feet. They astounded me by scaling some sheer slopes and leaping when they had to. Once they hit the water, however, they were like sleek bullets, there one minute, gone the next.

I learnt that one of the last land colonies of these birds was right here in Simons Town, my home. They appeared a few decades ago and now, a few hundred pairs live and nest here on this small stretch of the coast. It is possible they used to be here in years past and vanished once the settlers came in and started hunting everything in sight, but now they are back and depend on the goodwill of this town to keep their numbers up. Unlike in Kokkare Bellur, Simons Town is not a small village but a trendy seaside town with a growing urban human population. Again, unlike in Kokkare Bellur, the presence of the penguin brings in tens of thousands of tourists who help the economy of this little town thrive.

The African penguin, also called the black footed penguin, is classified as endangered by the International Union for the Conservation of Nature or the IUCN. In the early 1900s, there were close to 4 million penguins. Today, there are less than 55,000 birds left. In the last few decades, there has been an accelerated decline in the numbers and if this decline continues, it is predicted that the African penguin will go extinct in less than twenty years.

To learn more about these birds and why their decline in numbers was so drastic, I went to SANCCOB, or the South African Foundation for the Conservation of Coastal Birds. They are a non profit facility that specialises in the rescue, research and rehabilitation of sea birds,

*Penguins moulting. They shed their feathers from time to time to
grow fresh new waterproof feathers.*
Photo credit: Craig Foster

mainly penguins. The day I got to SANCCOB, I arrived in time to watch a little blue (or sub-adult) penguin without his sea feathers yet being brought in from the beach. He had been bitten on the head by a dog. The volunteers along with the doctor were cleaning him up and giving him shots. In the infirmary, there were other injured penguins. Most of them had been bitten by seals. We saw one who was on the road to recovery after having a massive bite wound on his stomach. The early photographs of the wound were quite graphic and I could not believe that he had survived. Sister Vannessa Strauss who runs this facility told me that young seals often attacked the penguins by biting their stomach because that is where all the fish the penguins swallow is. She also told me that dogs who run around on the beaches without leashes often attack the birds. Vannessa spends all her time here in SANCCOB and has dedicated countless hours fighting for and looking after the birds.

I learnt from her that the main reason for the drastic decline in numbers from the early 1900s was the human interference with the birds. Penguin guano or their excrement was highly prized as fertiliser. People started to collect this from all the nesting sites. What no one realised was that the penguins desperately needed their guano. The guano makes the soil softer and thereby easier for digging. African penguins make burrows in the ground for their nests and this also keeps them safe from land predators and the harsh sun. These nests are especially important when they lay their eggs as one parent has to stay with the eggs at all times and when the eggs hatch, the chicks need a lot of protection. By removing the guano for human use, we took away the one thing that made the soil softer for these birds. Penguins do not have paws and claws with which to dig. They use their beaks. Without soft soil, it becomes too hard to dig deep and the nests were becoming shallow and unstable, as dry earth cracks and breaks and caves in faster. People also started collecting the penguin eggs for food and this double whammy of weak nests and lost eggs ultimately decimated population numbers.

It was only in the mid-twentieth century that it was made illegal for anyone to either remove the guano or the eggs. It looked like maybe this would help population numbers grow but another huge threat was looming and today, that threat is only bigger and has done more damage than all the guano and egg collecting. That threat is fishing. The African penguin is a specialist feeder. This means that they eat a very specific diet and without the food that they eat, they will die. They will not adapt to eating other food as they are physically incapable of doing so. Their main diet is small pelagic fish. Pelagic fish are those species of fish that inhabit the water column not near the bottom or the shore of coasts, open oceans and lakes. The African penguins feed primarily on pilchards, sardines, horse mackerel and round herrings. They can also subsist on squid and crustaceans. While in some regions, they have changed their diets slightly to cope with the massive exploitation of their environment, they do not thrive outside of their regular diets. When pilchard numbers collapsed due to extensive fishing, the birds started to eat anchovies. While they can exist on these fish, the lower fat and protein content in the anchovies are not ideal nutritionally for the birds' long term survival. Commercial fishing is putting enormous pressure on these fish and thereby forcing penguins to swim further and longer to look for food. In the breeding season which, with these birds can sometimes be almost all year round, the pressure for finding food is much greater. One parent stays with the chick while the other goes out to look for food. They need to bring food back and regurgitate it for the chicks. If one parents stays out too long in the water and is exhausted and has not found enough food, the whole family suffers.

Just like storks and pelicans, penguins tend to stay monogamous for several seasons. Often, one breeding pair will use the same nesting site for several seasons. Male and female penguins look exactly alike. Each individual penguin is distinguished from another by the numbers of dots on its chest and the patterns of those dots. They reach sexual maturity at age three to four and live on average until age twenty.

Today, the African penguin breeding range extends from Hollams Bird Island off central Namibia to Bird Island in Algoa Bay. Non breeding birds disperse from southern Angola to Kwazulu Natal. There are twenty-seven extant breeding colonies, eight islands and one mainland site along southern Namibia and ten islands and two mainland sites along the coast of Western Cape in South Africa, of which one is where I live and six islands in Algoa Bay in Eastern Cape. They do not breed anymore in ten sites where they have once been recorded as doing so. With growing human populations and environment degradation and climate change, the birds are being forced into smaller and smaller areas and have to compete with cape fur seals and other oceanic birds for their survival. Colonies are also getting crowded.

It is estimated today that there are about 5,000 breeding pairs in Namibia and 21,000 in South Africa. The birds that nest near my home face predation from cats, mongoose, seagulls, ibis and dogs. Rules have been put in place to not allow dogs to run around without leashes but the enforcement is poor. The other mainland site which is in Betty's Bay, is away from the town and is even predated upon by leopards.

One evening, while walking along the coast in front of my house, I stopped to watch the penguins come in after a hard day in the ocean. The birds come in together in large numbers for safety. These groups are called rafts. Before they make a run for the shore, they sit on the ocean surface to monitor their surroundings. Once they feel that they are free of danger, they dive and bullet into the shore, sometimes using the waves to surf their way onto the rocks to get their nesting sites. On stormy days, it is much harder and they do face the danger of hitting the rocks wrong. As I watched several groups come in, I could also see some seals patrolling the area and right then, in front of me, one seal just bulleted into a group and came out with a penguin clutched in his mouth. Predation is sometimes one of the hardest things to see in the animal world. One part of me had been rooting for the safe return of the penguins while another could not help but admire the skill and patience it took for the seal to grab a bird. This is also specialised hunting behaviour. It is not the penguin in particular that the seal wants but rather the fish in the penguin's guts and that is why they wait to attack them when the birds come back from foraging in the ocean. I could not help but wonder if that particular penguin was a parent and if her partner was waiting for her return with hungry chicks. The seal thrashed around with the penguin and soon there was nothing left to see.

If one of a breeding pair dies it means there is no hope for the chicks if they are still young and unable to swim. This has given rise to a peculiar protection quandary and conservation politics. The sea birds people want isolated nesting and breeding sites for the birds and want the seals to be kept away. In fact, on some islands, rangers are employed to chase away seals who come on to the island, in order to keep the islands exclusive for the birds. The seal conservationists, on their part, find this a bad practice and many arguments have ended in accusations and hurt feelings. In the early years prior to the advent of huge numbers of people, hunting and collecting seals, guano and eggs, there were millions of birds and seals all living together. The occasional predation made no dent in numbers and life went on. Today with penguins classified as endangered and their days numbered if protection is not stepped up, the loss of every single adult breeder is huge. In this mess, one cannot argue, even though some seal people would like to, that seal numbers are up at about 2 million animals and quite stable unlike penguins and other sea birds like gannets and albatrosses and some cormorants. However, it is also quite certain that there are very few seal attacks as these birds are not a significant part of the seal diet and it is usually a few juveniles who attack the birds.

It is one of the eccentricities of modern conservation methods that a few species are favoured over others and get all the attention. This does tend to skew where the money goes, how it is used and if we are doing the right thing at all. Penguins and seals co-existed for millennia without people and once we arrived and literally and figuratively stuck our oars in it, it has been a mess.

On the beach in front of me some of the penguins are standing around looking irritated and miserable, their feathers falling off their bodies and flying around. This is the process of moulting and occurs once every year. The birds moult in order to grow fresh waterproof feathers. Prior to moulting, they gain weight and store fat. While they are moulting, for a period of twenty days they cannot feed as they cannot go to sea and lose up to half their body weight. Once they have moulted, they spend up to six weeks in the ocean feeding. While they are moulting, they cannot breed as they cannot go into the water to either feed themselves or their young. In the last few years, Vannessa has noticed an increasing number of chicks being abandoned as they have been born while their parents have gone into a moult or just being abandoned because food is so scarce and the adults can hardly feed themselves. SANCCOB steps in only in the most dire circumstances and takes in the chicks and then hand raises them. A University of

A female olive ridley turtle just after laying her eggs.
Photo credit: Kalyan Verma

The baby seal who was stranded after a storm. It took me a few hours of sitting with her before
she would let me touch her and then help her back to the ocean.
Photo credit: Craig Foster

On a boat on the Chobe river, tracking elephants and crocodiles.
Photo credit: Craig Foster

A gharial in the Chambal river.
Photo credit: Kalyan Verma

Gharials basking in the sun. As they are cold-blooded they like lying in the sun to warm up their systems.
Photo credit: Kalyan Verma

The Rann of Kutch. The succulent grasses hold water here.
Photo credit: Kalyan Verma

Wild asses in the Rann of Kutch.
Photo credit: Kalyan Verma

The dry Savuti grasslands in Botswana. The waters will soon fill these plains.
Photo credit: Craig Foster

The big cat fish pool. The day before the waters from the Angolan Highlands arrived in the Channel.
Photo credit: Craig Foster

The waters from the Angolan highlands filling the Savuti channel, heading towards the cat fish pool.
Photo credit: Craig Foster

The Savuti channel now flowing fully, the cat fish pool is entirely submerged.
Photo credit: Craig Foster

The tired cat fish I helped swim upstream once the waters came. After almost a year of just surviving she was too tired to swim upstream and was pushed towards the banks by the currents. I helped her get upstream.
Photo credit: Craig Foster

A penguin couple. Penguins mate for life.
Photo credit: Craig Foster

Cape Town study showed that birds raised by SANCCOB have a higher survival rate than the ones being raised by their parents. This could be because at SANCCOB an effort is made to feed the chicks the exact food they need to mature while in the wild today, penguin parents struggle with finding the right quantities of food and even the right quality of food.

One of the big and present dangers for penguins and indeed the other sea birds here are the oil spills. Many oil tankers make their way around the Cape on their way to the rest of the world and many of them leak oil. Oil slicks get carried across the ocean and if a sea bird gets coated in it, it is an immediate death sentence as the feathers become too heavy for them to fly or float and they lose their waterproof quality. A lot of oil on the birds can actually suffocate the bird to death. In 2000, in one of the biggest environmental disasters, 20,000 birds were involved in an oil spill. It took SANCCOB and hundreds of volunteers from around the world three months to get the birds cleaned. Another spill of this magnitude will spell disaster for the penguin populations. SANCCOB gets no support from the government for its work with penguins and other sea birds. They are, however, the only place that looks after these birds when they are sick or injured. Millions of rands are made every year by SANPARKS (South African National Parks), in Boulders, the protected sanctuary for the penguins in Simons Town, but none of this money goes to SANCCOB. It is rather ironic that it is the birds who make the money for the government and yet do not directly benefit from this money they make.

SANCCOB survives through fund raising and private donations. At SANCCOB, at any given time, there are dozens of volunteers trained to deal with the birds. They are also always prepared and on standby for any huge disaster like an oil spill or giant storms that might orphan chicks or injure birds. The one giant plus is that penguins can be hand raised and released and they still stay wary of people. The African penguin might look small and cute and many visitors assume that they might like being touched. This is a bad move as the birds are wary and mistrustful of people and can bite quite viciously.

I attempted to learn to feed a few baby penguins from Vannessa. She showed me how to pin them down upright between my legs, tilt their heads back, use two of my fingers to open their beaks and then push a small fish down their throats. The baby birds can only be fed like this as in nature, they will tilt their head back and open their mouths while their parents regurgitate the food directly into their throats. Fuzzy and fat, the babies are unbelievably cute, but I learnt very quickly that they are super strong and can bite quite nearly as hard as their parents. It took many tries to get one fish down one baby bird's throat. Chicks in nature fledge in a period of 60 to 130 days. This wide

Penguins are essentially sea birds.
Photo credit: Craig Foster

range of days is determined by how much nutrition and care the birds receive. Once they are fledged, they are a steel grey blue colour and are called 'the blues'. These birds, once fully fledged, can take to the ocean. These juveniles go to sea on their own and come back to their natal colony after a lengthy time period of 12 to 22 months to moult into adult plumage. It's only the breeding pairs that spend a lot of time on land, while the others spend most of their time in the ocean.

African penguins are not very big. Their full height is about 70 cm. But their body is like a sleek giant muscle and in the water, they are like speeding bullets. The black-and-white colouring enables two things. One, any predator looking up from the sea bed will find it hard to spot the birds because of their white underside and, two, predators from top will find it hard to spot the birds because of their black top side.

Watching Vannessa and the volunteers hard at work in SANCCOB, I felt such respect for what they were doing. Just like Manu in Kokkare Bellur, these are people who care and have a passion. They are not doing this for money or fame; they are doing it because they believe that these birds need a fighting chance to survive.

In Simons Town, homes next to the penguin colony have the birds sometimes nesting in their gardens, we have them crossing the road, and several do die in road accidents. They are loud birds with a braying sound, hence their other name – the jackass penguins. Cars parked outside on the road always have to be checked for birds sitting under them and many rules and regulations on how to access the beach where they are found and the areas around their nesting sites do cause some problems for the residents. The biggest problem, though, is not the birds but the tourists who come to see them. Eager to make money every season, there is not much regulation of the traffic or where people park or even how they behave. It cannot, however, be denied that the tourists have given Simons Town an international profile and the little town's economy is quite dependant on them. Even without the penguins, tourists will come here as Simons Town is a naval base and a quaint seaside town with excellent beaches and is the closest town to the Table Mountain National Park, but the numbers will be far less. To me it feels like this is the rent the birds pay for some of the best real estate in the world…a pristine stunning beach.

A Spot-Billed Pelican. They come to Kokkare Bellur every year during the monsoon.
Photo credit: Kalyan Verma

Here too, the birds over decades have chosen to nest and stay in the heart of a human community. Even though these birds are nervous around humans and do not like us coming too close to their burrows and nesting sites or even to them when they are resting on rocks and on the beach, they stay. Part of the reason could be that here the beach is more protected from massive storms that lash other beaches on the coast, less exposed to the open ocean and it's easier for the birds to launch out to sea and return here. I would like to think that one more reason would be that they trust us, the people who live around them and ensure that this area is a preserved as a penguin sanctuary, where they can nest and have their chicks in peace.

And maybe it's time to say a big thank you because what the penguins, the storks and pelicans bring to us is the ability to think beyond just ourselves and our lives. They give us the opportunity to step outside of our selfish little constructs and make us acknowledge that this world belongs to other fellow beings. I give them right of way when they cross the road, I watch for them at night and drive slowly, I do not walk where I know they have their burrows and I walk around them if they are resting on a rock. I ask for people to leash their dogs, I ask people to not litter on their beaches. If I find one injured I take her to SANCCOB. I continue to support the endeavours that make this coast a marine no-take zone, hoping that this will give the pelagic fish these birds survive on a fighting chance against commercial fishing. When I watch them waddling, swaying from side to side as they walk looking like escaped nuns, I feel my day's stresses, tensions, negativity all just slip away. I find my mouth curving into an involuntary smile and for that moment, in that moment, I have nothing but the penguins in my sight, the sea air on my lips and the sky above me. Perhaps this is their real gift to us when they choose to come and live where we are. The gift that allows us to step outside the high chaos of the lives we have created and for a moment, just feel the living world around us.

> *Humankind has not woven the web of life.*
> *We are but one thread within it.*
> *Whatever we do to the web, we do to ourselves.*
> *All things are bound together.*
> *All things connect.*
>
> — Chief Seattle, 1854

7

OLIVE RIDLEYS
AND SHARKS
HEART OF THE OCEAN

Take a walk with a turtle, behold the world in pause.
— Bruce Feller

MISTY RAIN fell in a steady rhythm and I kept wiping my eyes as I struggled to keep up with uncle Siddharth striding along the coast

line on Besant Nagar beach, Chennai. Foam from the waves coming in lapped around my feet and the whole world was a pearly grey. I walked behind him, and beside my small frame he seemed like a giant, his long arms swinging with energy and his silver hair waving in the breeze. He stopped suddenly and looked up towards the beach and then motioned me forward. From where he was standing, strange tracks led up the beach. The sun was still struggling to come up, blanketed as it was by a grey cloud. The tracks were wide apart and something I had never seen before. We followed the tracks and they seemed to end abruptly on a mound of sand. I looked around but could not see anything else. Uncle just smiled and said to me that we were standing near a nest of turtle eggs. I could not see any nest, let alone eggs.

'It's underground,' he said, 'buried in the sand. A sea turtle came up here and laid her eggs.' Then after a pause, he muttered to himself, 'I think an olive ridley. The tracks do not look wide enough to be a leatherback, leatherbacks are more rare.' I stared up at him, my eight-year-old mind intrigued at the idea of a big sea turtle. He went on to tell me all about sea turtles. Years later, I would stand on a beach myself and watch over ten thousand turtles dig their nests and lay their eggs and his words would play in my ears.

Sand was flying in every direction. Some of it was in my mouth as I knelt in awe near a mother turtle and watched her dig her nest. Her big round body swaying as her flippers dug up sand. From behind her, I could see the hole she had dug was quite deep. It was amazing to think that she could dig such a wide and deep hole into the sand with just her flippers. The walls of the hole were hard-packed sand and the whole nest looked like a pot that was buried into the ground. Then I watched as she laid at least 100 eggs. Fat tears rolled down her cheeks. I stroked her smooth head and laid my hand on her carapace. The shell around my hand lit up. Phosphorescence. The sea of her distant travels had come with her. It was like magic. All of it. This ancient wise being and me on a wild beach. I was in Orissa, in the Bhitarkanika Sanctuary, on Gahirmatha beach, a sanctuary especially created for the turtles and their nesting. I was here to film this scene for *Born Wild*. It was, luckily, an unusually calm sea, otherwise getting to this beach can be really hard. The ocean here is vast, wild and fierce. That night, however, it was calmer, allowing us to get on to the beach.

In the slap of the surf and the cry of the wind I could hear uncle Siddharth's voice.

Turtles are the oldest living reptiles on the planet. Recent fossil records have dated them as being over 220 million years old. Sea turtles are marine reptiles that live most of their lives in the ocean. The male turtle hatches on land but once he returns to the water, he never comes back on land again. Female turtles, on the other hand, come up on land once they reach their sexual maturity and are ready to lay eggs. Female turtles come back to the beaches where they were born. There is a homing device built into their brain that guides them back to the beaches of their birth, a natural technology far more sophisticated than anything we have ever invented. There are eight species of sea turtles, and nearly all of them are threatened or endangered. The olive ridley turtle is classified as a Schedule 1 animal in India and that means they are meant to be as protected as the tiger.

Of the eight turtle species, only the olive ridley and the Kemp's ridley mass nest. They gather in massive numbers and come up on to the beach to nest. Solitary nesting has also been observed in over forty beaches, but their mass gathering in six beaches of the world in huge numbers makes them unique. This gathering is called the arribada, as it was first studied in Mexico. The word 'arribada' means arrival.

The largest arribada of olive ridleys in the world is in Gahirmatha, in Orissa.

Turtles mate in the open ocean when the females reach sexual maturity and are ready with eggs. The females are usually about fifteen years old when they reach sexual maturity. The natural lifespan of a sea turtle is about sixty to eighty years.

As I sat near the female turtle and watched her rock from side to side to pat the sand down over her nest to flatten the top and hide her nest, I wondered how far she had swum away from this beach where she was born and what she had seen. I remember uncle Siddharth telling me, 'These are some of the oldest creatures on the planet. They are special because they are born on land but live in water. Not too many creatures like that eh?' Then his voice had turned soft and pensive. 'These ancients, Swati, imagine what they have seen? They were witness to dinosaurs turning into birds and they also must have witnessed the first mammal turn into elephants, whales, bats. They probably witnessed the leap from ape to humanids and then watched the first Homo sapien emerge.'

The olive ridley is a pelagic sea turtle, although some are found in estuaries and coastal waters. Their nesting season occurs over several months a year, and in that time they might come up to the beach to nest over five times. To do this, they make a journey from their pelagic holds to coastal waters to nesting beaches and back. Ridley turtles have been observed at 4,000 km out to sea. Their migration patterns, where they go after they hatch, how they live, survive and so on, is still a big mystery with just general information available on their behaviour and feeding.

They eat algae, sea weed, lobsters, shrimps, crabs, tunicates, molluscs and fish. They are also deep sea divers, getting down to more than 150 metres for benthic invertebrates or animals found at the bottom of the ocean. The leatherback turtle dives the deepest to over 1,000 metres.

The olive ridley turtle is considered the most abundant sea turtle in the world with nesting populations estimated at about 800,000 females. These turtles have a wide distribution in the South Atlantic, Pacific and Indian oceans.

However, their populations have crashed by over 50 per cent since the 1960s. In the last decade, there has been a noticeable drop in numbers of turtles nesting at the Gahirmatha beach in Orissa. In the period 2000–2010, over 100,000 turtles were recorded as dead. Over ten thousand turtles died every season. Worldwide, commercial fishing practices are the main reason for this. Commercial trawlers, utilising gill nets, purse seines, long lines and drag nets. I have seen large breakaway drift nets with hundreds of entangled pregnant female turtles all dead on the beach. The turtles get stuck in the nets with no way to get out and drown. The trawl nets are designed to drag the sea bed to catch shrimp and in the drag, often catch any other animal unlucky enough to be in its path. This is one of the most destructive forms of fishing with more than 11 million metric tonnes of bycatch being the collateral damage. Dolphins, sea birds, whales, sharks, seals and many other animals are trapped by these nets. Big animals like turtles can't get out of the way fast enough and are trapped.

While turtles can hold their breaths for a long time, they do need to come to the surface to breathe air; unlike fish, they have no way to breathe under water. Nets might stay in the water for hours, killing them. Even when the net is brought up while the turtle is still breathing, the rapid ascent of the net causes decompression sickness in the animal. This is when nitrogen bubbles enter the bloodstream. This happens as under water, the pressure is very different from that on the surface, and a rapid ascent alters the pressure too quickly. Human divers with scuba equipment who surface too fast suffer from this and it is suspected that turtles do too. In humans, it causes extreme pain, disorientation, paralysis and death. Not enough study has been done to see what the symptoms might be in turtles.

With thousands of these turtles washing up with the nets, the Indian government made it mandatory for trawlers to use a TED or the Turtle Exclusion Device. This is a large gate-like structure built in to the net that allows large animals like the turtle to find their way out of the net. International tests have shown that these devices are 97 per cent effective in preventing turtle deaths. There is a small loss of catch between 5 and 13 per cent, and though this data has been scientifically proven, many trawler owners and shrimp fishery owners have claimed that the loss is as high as 30 per cent and refuse to install the devices. In India, despite the law, very few trawlers have bothered to outfit themselves with the device and there is hardly any enforcement of the

Olive Ridley turtles coming to the beach to nest.
The female turtles come back to the beach where they were born to nest.
Photo credit: Kalyan Verma

law. Trawlers are also asked to maintain a 5 kms distance from the shore during the nesting season. The actual nesting areas on the islands of Gahirmatha and Rishikulya and Devi river mouth, are no fish zones. Even this is not monitored. The Orissa government, on being questioned, told us then that it was impossible for them to monitor all the trawlers or even to patrol the beaches. However, activists showed us proof of cash paid to the forest department for two sleek and fast boats for just this purpose. We got no answer from the government regarding this. The few years that the navy was enlisted to monitor the beaches made it much safer for the turtles but it is not the navy's job. Many of the trawlers in Orissa's waters are also foreign boats from Thailand, Indonesia and other countries and have no right to fish in these waters. Local fishermen have often complained about this but don't do much as checking the trawlers will lead to all trawlers being examined.

Often, when people I meet in the city hear my stories, they will ask me why the turtles are not being better protected and they will blame the fisheries. Then I tell them that the plate of prawns they just ordered helped kill some turtles. This is not an answer anyone wants to hear. But I firmly believe the only way to ensure that all trawlers use the TEDs is if consumers will only eat shrimp caught ethically without the damage of bycatch.

In one of the greatest tragedies of the oceans, often in big seas or storms, the nets break away from the boats and drift along the sea. These nets, called ghost nets, fish for the entire span of their lives, which might be years and years until they wash ashore. These ghost nets entangle hundreds of thousands of marine animals including turtles and keep killing with nothing and no one to stop them. Even when they wash up on shore they can be dangerous, trapping shore bird chicks and adults in the sand. This is called ghost fishing, and it is a major problem.

Plastic in the ocean is another problem. A lot of the garbage we discard without a thought ends up in our water systems and makes its way to our oceans. As plastic is non-biodegradable for a long time, it can stay viable in the ocean. Turtles eat jelly fish and a floating plastic bag often looks like one, so this has also become a major problem. Even when the bags disintegrate, the smaller bits are eaten by fish, causing damage. While it is easy to want to blame the fishermen for all turtle deaths, I think excusing ourselves from the issue is not just naïve but dangerous.

Habitat loss is another problem. Many nesting beaches no longer function as nesting beaches simply because they have been taken over by development activities. Sea turtles are phototactic, or sensitive to light. They can navigate by starlight and moonlight but any other source of bright light confuses them, especially when they are hatchlings. With coastal developments growing and roads and highways approaching beaches and with bright home and street lights, many turtles, both young and old, are found walking towards those light sources instead of the ocean. On many nesting beaches, especially the beaches that host the arribada, people also dig up the eggs for food and even the adult turtles are consumed for both meat and leather. This was especially a problem in Costa Rica and Mexico, decimating turtle numbers. When these turtles are confined to smaller and smaller patches for their arribada, they have a tendency to dig over and dig up each others' nests, destroying their own eggs. It is natural for a certain amount of eggs to get destroyed like this which is why a single female will come up to nest at least two if not three times in the season. But when their habitat shrinks and eggs are being removed by people, every egg is a loss. On one beach in Costa Rica, a successful programme with local people involved egg collection only from the first wave of turtles as those nests are often dug up by the turtles themselves. The rest of the nesting season, the locals patrol the area, ensuring no one digs up the eggs and also making sure that stray dogs and other predators do not dig up the nests. They also wait for the eggs to hatch and help the hatchlings get to the water.

In Orissa, too, efforts are being made to safeguard the eggs and help the hatchlings to the ocean.

When a small baby turtle struggles out of the sand and then has to make its way to the ocean, it is one of the most poignant sights. These babies are so small that they will fit in the palm of your hand with room left over, using their little legs to propel themselves towards the water. As they crawl as fast as they can, wheeling and diving gulls, crows, brahmini kites and other birds just pick them off. It's heartbreaking to watch but it is the cycle of life and it has played out like this for millions of years. When one little baby finally makes it to the ocean, it is by no means the end of its ordeal. There are big waves, currents, and predators waiting for them there too. When we see one adult turtle, we must respect what she has faced and how masterful she is to have overcome all her struggles and survived to her age.

The eggs have a two-month incubation period. The hatchlings hatch together and this process can take several days. They all need to hatch together to help each other emerge from the nest. The broken eggs act as a base from which they can climb up and the turtles on top scratch their way out of the sand. When they emerge en masse, they also increase their chance of survival because in large numbers, they can overwhelm a waiting predator and at least some get away. They then climb out of their nest and make their way to the ocean. There is an in-built instinct that makes them seek the brightest light source in front of them and on any given day on a dark beach, that source of light is the horizon, so the babies move towards the ocean. It is estimated that 1 in roughly 1,000 survive. In order to increase the odds, especially now with populated beaches and disturbed nesting sites, conservation organisations build guards around nests and then help the hatchlings into the ocean. But once in the water, the baby still has to survive the vast world of blue to adulthood. While buried deep in the sand in the egg, the female hatchlings imprint on the beach as this is where they will mostly come back to nest when they reach adulthood.

Recent studies have indicated that sea turtles can detect the intensity and the angle of the earth's magnetic field. Using this ability, turtles may be able to find latitude and longitude, enabling them to navigate anywhere. This is being further investigated. It is also widely believed by experts that the hatchlings imprint the unique qualities of their natal beach when they make the run from the nest to the ocean. This might include smells, topography, low frequency magnetic waves and, maybe, even the ocean currents. It is like they have a magic thread, an energy beam that always keeps them connected to the beach of their birth no matter how far they swim away.

Turtles are not the only reptiles who return to where they were born. Crocodiles, too, return to their original nesting sites.

I remember the first time I heard this whole story from uncle Siddharth, I was amazed, but then years later, when I saw hundreds of dead female turtles, I was horrified. As humans, can we even understand what it takes for each one of these animals to survive to adulthood, become sexually mature and then come back to the beach where they were born, just to die? And the death of every egg-bearing female is a further hammer blow to an already vulnerable population.

Now climate change is also playing a role in the nesting as sand temperatures determine the sex ratio. The hotter it is, the eggs produce females and the colder it gets, they produce males. There is an optimum temperature at which the ratio evens out between males and females. A turn in the ratio in any one direction will also challenge the future survival of these creatures.

And that survival is imperative. Sea turtles are a keystone species. Keystone species are a species with a large sphere of influence. Its presence and behaviour help determine what species in what number will make up an ecological community. Removal of a keystone species creates a domino effect that resounds through the entire ecosystem, slowly damaging biodiversity and endangering the entire survival of that ecosystem.

So if sea turtles go extinct, there will be several ecological effects. Sea turtles, especially the green turtles, are one of the few animals in the marine environment who eat sea grass. Just like grass on your lawn, sea grass too needs to be cut constantly for it to grow well and thrive. Sea grass beds are important because they provide a safe breeding and developmental ground for many fish, shellfish and crustaceans. Over the past decade, there has been a decline in sea grass beds and this is having a knock-on effect on species that use them as breeding and developmental grounds. There is another very valuable service provided by the sea grass. Sea grass sequesters huge amounts of carbon. A recent study has found that sea grass sequesters nearly as much carbon as forests do on land. With climate change and our ever growing carbon emissions, the ocean has had to bear the burden of sequestering vast quantities of carbon. This is making the water more acidic and creating major problems. The study found that the global carbon pool in sea grass beds was 19.9 billion metric tonnes. The research also showed that though sea grass occupies 0.2 per cent of the ocean beds, it sequesters almost 10 per cent of the carbon. If sea grass were to disappear, it would be a catastrophe for the ocean and for us.

It's not just the marine ecosystem that sea turtles influence but also the beaches and dune systems. Normally, beaches and dunes do not retain nutrition very well because of the sand. So very little vegetation grows on the dunes and hardly any on the beach. Sea turtles, on average, lay hundreds of thousands of eggs in one nesting season. Not all these eggs hatch, and not all the hatchlings survive either. The unhatched eggs and trapped hatchlings in the sand become a good source of nutrients for the dune vegetation. On beaches where turtles nest, the dune vegetation is able to thrive and become stronger because of the turtles and this in turn enables the stronger vegetation and root systems to hold the sand and check erosion.

Really, the big question seems to be, can an ancient creature that survived catastrophes that wiped out the dinosaurs and so many other species on earth, survive us? Or perhaps, the better question is can we survive the damage we are inflicting?

And not only are we hurting the marine ecosystem by decimating habitats, polluting seas and killing turtles but we are also killing the apex predators of the oceans. We are destroying two keystone species and if we continue down this path, there is no coming back from this.

A gully shark in the waters of False Bay.
Photo credit: Craig Foster

*Sharks are beautiful animals and if you are lucky enough to see
lots of them, that means you are in a healthy ocean. You should be
afraid if you are in the ocean and don't see sharks.*
                                                        – Dr Sylvia Earl

One hot summer morning over thirty-five years ago, around the
time I saw the sea turtle tracks on Elliots beach in Chennai, along
with uncle Siddharth, I watched fishermen bring in their haul. It was
mostly small fish but tangled in the net was a medium-sized shark.
I watched as they removed all the fish, making piles, and then they
detangled the shark and chopped its fin off. It is one of the few times
I have seen uncle Siddharth furious and I watched as he strode over to
the fishermen and had an argument with them. I caught the gist of it
with the fishermen saying that the shark got entangled in their net by
accident and they could not set him free without losing their catch, but
now that they had him they were going to take the fin. They said that
there were many boats that did this. Uncle came back to me, he looked
pensive. We were, ironically, standing near a monument on the beach
built for a boy who had drowned in the ocean. Local word had it that
he had been attacked by a shark. 'I swim in these waters every day,'
Uncle said. 'I have never seen a shark but I know they are there. What
does it say about us a species, Swati,' he said, 'if we kill everything we
are afraid of?'

'And this worries me, this shark fin business. What a waste.
These are predators, Swati, as much as the lions or tigers on land,
they are the lions and tigers of the ocean. At least, if the whole animal
is consumed, it's one thing, but just for this fin?' He shook his head.
'Apex predators are very important to life on earth...by taking life, they
give it,' he said.

Those were big words, it was a big idea, but I would experience
and learn what he meant.

Years after that long ago summer morning, I was on a boat heading
towards Seal Island, in False Bay off the coast of Cape Town. I now live
here and I was eager to see the most well-known denizens of this bay,
the great white sharks. The ocean and the sky blended together in a
crystal clear blue, dazzling the eye. The sounds of seals filled the air as
I stepped over the side of the boat into a cage. As the water closed in

over my head, I forgot to feel cold or even to breathe because around my cage were about eleven sharks.

It is not until you are in the water, eyeball to eyeball with them, that you actually appreciate just how large they are. These were adult sharks, about 3 to 4 metres in length and each weighing between 800 pounds to a ton. A second revelation was that in the water they looked calm and beautiful, not the scary monsters from our dreams, photographs and films.

There was only one shark amongst the eleven who felt a bit tricky. He was called 'the nutter' and had an edge to the way he swam around us. Craig, my husband, a wildlife and anthropology filmmaker, who was with me, said he would have been fine in open water with all of them, except the nutter. I asked him why and he said it was because the nutter displayed subtle signs of warning, as if saying 'keep away'. He was also swimming a lot faster and with more purpose than the others who were cruising around more out of curiosity.

One of the more obvious displays of 'keep away' was a slight gaping the nutter did with his mouth. This is a huge sign of aggression and irritation in these animals.

Craig has spent hours open water diving and filming with great whites and tiger sharks and he has only ever had memorable experiences. With the tiger sharks, after weeks of being in the water with the same individuals, he was riding on their backs holding on to their fins. Great whites and tigers are two of the three shark species listed as most dangerous to humans. For all these dives, the crew had to bait the sharks in. Just going out into the ocean is no guarantee of spotting sharks. Even baiting sometimes takes hours before the sharks come in.

Craig has been swimming in the ocean all his life. In the last four years, he has been in the water nearly every day. Of course, he swims without bait. In all those thousands of dives, he has seen the great white swim by once. The second time he saw one was while he was on his kayak. Other people have swum all their lives and never seen them.

I see seals, dolphins and whales all the time in the water. I have even seen killer whales in the water but until I went cage diving, I had never seen the great white even though they are year-round residents of False Bay.

To see them when I went cage diving, the dive conductors had to use chum to bring them in. They tossed fish blood in the water to attract the sharks. Yes, many sharks have an acutely strong sense of smell. They can smell one part of blood in millions of part of water. Now, do not get excited by that statistic. This means that they can smell a drop of blood in an Olympic pool-sized area. So if you are swimming and for some reason, you have cut yourself and are bleeding, a shark might come into the area if he is already in that area. Hordes of them will not be making a beeline to you. Experiments have proven that human blood and cow blood have not attracted sharks while small amounts of fatty fish blood brings them in quickly.

I am not saying they are tame. They are efficient, amazing predators designed to be the apex predators of the oceans and a whole lot of respect is their due. For over 400 million years, they have contributed to evolution in the ocean. Every fish on their menu has evolved to become stronger, faster, bigger, in an effort to outwit this amazing predator. Sharks also take out the infirm, the weak, the sick and the old. They also scavenge on the dead. What this means is that they keep the oceans healthy and clean.

On an average, there are five shark attacks on people a year. Many of them non-fatal. It also seems that certain kinds of people attract

Ridley turtles on the beach for the mass nesting.
Photo credit: Kalyan Verma

them. In a sea full of people, sharks have been known to swim past hundreds to get at one. No one knows why these attacks happen. There is no real way to explain it. Statistically, 90 per cent of the attacks are on men.

In the last few years, there have been some attacks on people on beaches just a few kilometres from where I live. This does not prevent me from going into the ocean. If I thought I was going to be attacked by a shark every time I went out into the ocean, I would be a super rich woman by now because I would have bought myself lottery tickets. It's a lot easier to win the lottery than it is to get eaten by a shark.

The fact that more people die driving to the beach on a daily basis compared to the odd fatal shark attack is drowned in the clamour of hysteria over a single attack. If there should be hysteria, it should be over the fact that experts believe that there are less than 3,500 great white sharks left in the world's waters. Basically, there are less great white sharks in the water than there are tigers on land and we all know how much attention that fact garners. So why are we not more concerned about the sharks?

In western Australia, the government has actually started to kill sharks of certain sizes on beaches that are used by swimmers and surfers. This was instated after a few surfers were attacked. In Durban in South Africa, there are shark nets in the water to prevent sharks from swimming into the shallower waters where most people swim. These shark nets often tangle up and kill sharks and dolphins as well. Imagine if governments banned all cars on roads because of the deaths caused by careless driving or banned all guns and alcohol. That will never happen, so why is it ok to kill sharks?

Diving with them in open water is a risk as it is a risk to interact with any predator. The astonishing fact is that you can get a lot closer to the shark than you can to a lion or tiger on land without being killed. The main prerequisites are very clear water, so both you and the shark can see each other coming. The other is to not dive with them in deep water. Deep water gives the predator the advantage as these big sharks come up vertically to attack.

It was only after I moved to South Africa, and started spending a little time in the water with Craig that I even realised that there were so many different kinds of sharks in the oceans. Of course, on an intellectual level, I knew that there were several shark species out there but it was a vague piece of knowledge and I never really thought about what it meant. What it means is that there are dozens and dozens of more shark species in the waters of the world that are small or medium sized. Altogether, there are 400 sharks species in the water, out of which exactly five have been known to occasionally attack humans. The great white, the tiger shark, the bull shark, the oceanic white tip and the blue shark. From the way sharks have been demonised and even glamourised by various media outlets, documentaries and films, one would think that the only sharks that exist are these large predators. When is the last time anyone saw a film on the pyjama shark or the striped cat shark, a 1 metre-long predator, that lives in kelp forests, that sleeps in caves and eats crabs and small fish? Or on the various shy sharks, that are about 60–80 cm long and, true to their name, try and stay hidden, or the leopard cat shark which is 80 cm long that also lives in the kelp? Or even the cow sharks, or the seven-gilled shark, a 3 metre-long fish that will hunt and kill dophins and seals and is yet to kill a person? Or the spotted gully sharks that are 2 metres long, that love to surf the waves and sometimes come together in numbers in excess of 200 in literally knee-deep waters and will just flee if a swimmer or diver makes a sudden movement in the water? The largest shark in the world is the whale shark, about 13 metres long, and the smallest is the dwarf shark that measures just about 6 inches.

Why do we have such a negative reaction to sharks? It cannot be only because of that one film, Jaws. Sure, Jaws demonised the great white shark in a way that lead to their mass slaughter, but almost forty years after the film, we can't still be reacting to it. If we are, then the reason is not enough has been done to counteract the negative message of the film. Most people I meet still just hear the word shark and equate it with a killing machine. Maybe it's also because being fish, they don't appeal to our mammalian brains the way dolphins and whales do. Of course, one could argue here that dolphins and whales have not killed people in the open ocean that we know of. Killer whales are big predating mammals of the dolphin family and they have killed several people while in captivity, but again, not in the open ocean. Could it really just be that? Sharks have killed people and that is enough for us to have built up this enormous irrational fear of them. By that logic, every human should run screaming from every other human. There are no greater killers of people than people and let's not forget mosquitoes. The average household pet dog has bitten more people than sharks. You are more likely to get killed by falling coconuts, lightning strikes, chairs and toasters, not to mention cars and people, than sharks. There are warm blooded sharks and cold blooded sharks and fresh water sharks, and we know very little about all of them. My 14-year-old son not only dives regularly with his father but has learnt to handle all the smaller sharks and swim with the seven-gilled sharks or the cow sharks. The seven gill cow shark is a large shark. They are about 3 metres long and are formidable predators. They hunt and kill seals and dolphins and other sharks. They hunt in packs and a group of them are found just a few kilometres down the coast from my house. My son has been diving with them since he was twelve years old and its one of his favourite things to do. He will never be able to interact with land predators in this way swimming close enough to touch them at times.

Research from the waters around Bimini islands in the Bahamas suggests that far from being killing machines, sharks have individual personalities. We only associate and attribute personalities to dolphins and whales, but scientists have, over the years, collected enough information on sharks, to prove that many of them are social, have complex social networks, some of them even have 'buddies', and each individual is either calm, or nervous, or aggressive or shy, with no one animal similar to the other. The scientists from the shark lab in Bimini, led by Dr Samuel Gruber, who also first looked at the phenomenon of tonic immobility in sharks, have also proved that they can be trained to perform tasks, and can retain information of that training for a long time. Tonic immobility is a state of stasis that the shark can go into when stimulated in a certain way. Small sharks can be put in tonic with gentle strokes, or just being flipped over on to their backs. Bigger sharks can be manipulated by stroking the side of their mouth where their ampullae of Lorenzini, a bundle of sensitive nerve endings reside. Sharks are highly sensitive to energy signatures in the water and the ampullae of Lorenzini are their detectors. Just like sea turtles, many sharks swim great distances to return to their spawning nurseries. They teach other where and how to find food and even when several of them are on a kill, they have an orderly hierarchical feeding system where they each take turns to feed. When all of this and more information that is being learnt everyday by observing them is brought together, it forces us to look at the shark differently and it should obliterate the irate irrational killer from movies.

I have had the wonderful experience of diving with several of the smaller sharks, like the striped cat shark, the leopard shark, the shy sharks and the spotted gully sharks. And watching Craig photograph and film them and spend time with them, I have also come to understand just how little we know about them. In Durban, the period between December 18th, 1957 and April 5th, 1958 is referred to as black December. In this period, there was an unusually high number of shark attacks on people. 9 people were attacked and 6 of them died. Analysis later showed that several factors had come together that December to create the perfect storm. For one there were a huge number of whaling vessels in the water and so blood and whale bits were everywhere attracting the sharks in and a huge flood had washed a lot of silt into the waters making them murky. The floods had also killed large numbers of livestock and all the carcasses too had washed down with the water run offs into the ocean. This period is also the peak tourism period in South Africa. The hysteria understandably at the time was high and tourists fled the beaches in droves causing a devastating attack on an economy that is highly dependent on income through these months. It affected tourism for a few years. In response to these attacks the Kwazulu Natal Sharks Board was formed in 1962. The Sharks Board authorized the placement of shark nets and drum lines along 320 km of coastline along the Kwazulu Natal province. These nets and drum lines were placed in order to prevent sharks from being able to swim into the shallower waters where most of the bathers spent their time. While it sounded like a good idea, the shark nets have been an absolute and complete disaster. They have done their job and protected the swimmers, but it could also be that without the shark nets people would have been safe anyway as none of the conditions of black summer have repeated themselves. The nets however have done nothing to protect the sharks. A shark net is a net that is submerged in the water to prevent sharks from being able to swim past them. Due to boating activity, the nets are submerged about 4 metres below the surface of the water and float above the bottom of the sea bed.

The sharks could technically swim above or below the nets. The nets are also staggered in the assumption that this will deter sharks from swimming past them. The nets are about 6 metres high and 214 metres long. A drumline consists of a large, anchored float from which a single baited hook is suspended. This baited hook is in place to catch sharks. It is assumed that these hooks are effective against the three shark species considered most dangerous to humans, the great white, the tiger and the bull shark. The nets were trapping sharks and the hooked line was trapping sharks. Hundreds of them die on this coast because of both. Its not just sharks that die but dolphins, turtles and whales too get caught by the nets. The hooked bait line often hook turtles as well. The sharks board is meant to monitor the nets and lines to ensure that they are functioning properly and to rescue any live animals caught by them. In the years the nets have been up and running over 35,000 sharks have been killed. Only about 155 of the sharks killed are the targeted three, of the whites, tigers and Zambezi sharks. The others were all sharks that pose absolutely no threat whatsoever to bathers. Horrifically, thousands of turtles, rays, dolphins and other sea creatures have been killed by these nets. It seems just a criminal waste to kill so many vulnerable animals in an already beleaguered ocean eco system just so that some people can feel safe while swimming. In another completely awful chapter of this horror story, the Sharks Board in Durban often hold shark dissections for the general public. They call this 'informing and teaching' the public and researchers. One can understand that dissections of dead animals for science might have some merit in terms of understanding them in detail but displaying that for public entertainement is truly macabre. It outlines our attitudes towards animals. If any government organization held the public dissection of a panda or even a lion or tiger, the public would lose their collective minds. We would all be outraged and storming the gates. However, a shark being cut up is business as usual.

One morning on the beach in my first week here in Cape Town, I came across a translucent brown pouch-like thing on the beach. It looked like an elongated seed pod and had curly stems on one end. Craig told me that it was called a 'mermaid's purse' and that it actually was a shark egg. It was the first time I actively thought about just how sharks give birth and how sharks are born. There are roughly two ways that sharks are born. In one, the mermaid's purse, or the egg, is laid and it attaches itself to its surroundings like coral, or kelp or seaweed or even the ocean bed with long curly stem-like tendrils. The yolk in the eggs feeds the growing shark embryo until they are ready to be born. There is no parental care involved and many eggs are targeted by a variety of predators. The successful birth of every shark and survival to adulthood is a hard tough road. Of the 400 species of sharks, roughly about 40 per cent of them give birth this way. This is called oviparity.

The rest are live births. The sharks stay inside their parent until they are fully developed. This is called viviparity. There are three ways this happens. Some of them have an egg case and yolk that feeds them and they are born when the embryo is fully developed and hatched from the egg inside the shark's uterus before being pushed out by their mother. This is ovoviparous. Some shark babies develop inside the uterus, and are not fed by yolk sacs but by other unfertilised eggs inside the mother, called oophagy, or eat their unborn siblings, embryophagy. A third way exists, which is the rarest. The egg's yolk sac becomes a placenta attached to the female's uterine wall and nutrients are transferred from the female to the pup. Yes, shark babies are called pups.

Just like the turtles, sharks are a keystone species as in the ocean, on the top of the various webs of life, they are the apex predators. They regulate the numbers of the prey, keeping a balance, not allowing any one fish or other animal population to become too big and throw the balance out of whack. For this reason, the prospect of a food chain minus its apex predators may mean the end of the line for many more species. A number of scientific studies demonstrate that a depletion of sharks results in the loss of commercially important fish and shellfish species down the food chain. Predatory sharks prey on the sick and the weak members of their prey populations, and some also scavenge the sea floor to feed on dead carcasses. By removing the sick and the weak, they prevent the spread of disease and prevent outbreaks that could be devastating. Preying on the weakest individuals also strengthens the gene pools of the prey species. Since the largest, strongest, and healthiest fish generally reproduce in greater numbers, the outcome is larger numbers of healthier fish. This is what uncle Siddharth meant when he said predators give life when they take it and that is why it

*A hatchling on the beach surrounded by a crowd watching
him/her make his/her way to the ocean.*
Photo credit: Kalyan Verma

is such a disservice when we make our apex predators look like just
simple killing machines in our films and books.

They also keep vital habitats alive by their sheer presence. When
sharks are present in the water, all other creatures flee. This means the
sharks apart from feeding on prey to keep the balance, also ensure
that their prey species do not indulge in behaviour that can destroy
habitats. For example, I earlier talked about how sea turtles kept sea
grass beds healthy, but if sea turtles overgrazed the sea grass, then
those habitats would be in danger again. Tiger sharks, which prey
on sea turtles, by their mere presence in some waters ensure that
not too many turtles stay in one spot, grazing the grass continually,
allowing for the sea grass to regenerate the way it is meant to. Both are
keystone species that are crucial to the health of the marine ecosystem
and they are so intimately interconnected. These are just some of the
connections we are so unaware of in our rush to exploit the ocean and
her resources.

Sharks are our apex predators and they have enjoyed an intimate
and crucial equation with ocean ecosystems for 450 million years. Can
we understand what that means? How critically entwined oceans and

sharks are? And here we are, destroying it at a rate where we might actually succeed in wiping out sharks in the next few decades.

And it's all just for the fin.

For just this fin, close to one hundred million sharks are slaughtered each year. Which means that 11,417 shark are killed an hour. Many shark species have reduced by over 90 per cent. We will witness the extinction of several shark species in our lifetime. Sharks are pulled up out of the ocean, their fins are lopped off and then they are simply flung back into the ocean. Often, the shark is alive, bleeding, and just sinks to the bottom of the sea bed as without its fins, it cannot swim. It's barbaric, unnecessary and a shocking waste. Majority of these fins get sent to the Far East where it is seen as a delicacy and status symbol. Marine biologists say that most shark populations have crashed by between 50–80 per cent due to this practice.

India is the second largest exporter of shark fins.

Which brings me back to that long ago summer day. Chennai city is one of the biggest hubs for the trade in fins out of India. Close to one million tonnes of fins are exported from here.

In 2013, the ministry of environment and forests sent out a circular to ministers and ministries in all coastal states expressly stating that 'any possession of shark fins that are not naturally attached to the body of the shark would amount to the hunting of a Schedule 1 species and thereby attract penal provisions under section 51 of the Act (Wildlife Protection Act) further in accordance with section 57 of the Act, the burden of proof for the unlawful possession custody control of such an animal, article meat, etc., shall lie on the accused.' This was done because without the whole shark, it is difficult to tell if the fins come from a protected species.

There are about 40–60 species of sharks found in Indian waters, of which ten species have been classified as critically endangered and are on Schedule 1 of the Wildlife Protection Act. Several other species are on Schedule 2, 3 and 4, making it illegal to hunt them as well.

The biggest problem is not the law then, but the will to protect and monitor these animals.

Sharks are not very fertile and mature slowly. This means they cannot replenish numbers as fast as we are killing them. Long line fishing, drum lines and large deep sea trawling nets are also killing them in large numbers.

It's a dangerous game of Russian roulette to play with the ocean. We have started a dangerous collapse that will only prove detrimental for our own long-term survival on this planet. The oceans are life-giving for various reasons, from being the primary food source for hundreds of millions of people living along the coasts to being the major producer of oxygen on the planet. Do we really want to mess with this? Marine biologists have warned that we have practically reached the tipping point in our oceans with plummeting fish stocks, islands of plastic garbage choking our waters, oil and chemicals and heavy metals contaminating the fish and us.

If our greatest challenge in this century lies in saving our oceans, we don't have a chance in hell of doing it unless we save our sharks, our turtles, and learn the true meaning of sustainable fishing.

It is a curious situation that the sea, from which life first arose, should now be threatened by the activities of one form of that life. But the sea, though changed in a sinister way, will continue to exist; the threat is, rather, to life itself.

– Rachel Carson, *The Sea Around Us*

8

CROCODILES AND SEALS
THE CRAWLY AND THE CUDDLY

Take the crocodile, for example, my favourite animal. There are twenty-three species. Seventeen of those species are rare or endangered. They're on the way out, no matter what anyone does or says, you know.

— Steve Irwin

ONE DAY, uncle Siddharth took me to the snake park, to watch a man called Romulus Whitaker do a show on snakes. He most nonchalantly picked up snake after snake, spoke about what they were, and went on to tell us in the audience why they were all magnificent, important and why our fears and superstitions around them were nonsense. He asked

for volunteers from the audience to come into the pit and touch the non venomous snakes to get an understanding of what they felt like. My father and uncle Siddharth encouraged me to go into the pit and feel it for myself. The snakes were beautiful, cold, smooth like silk and supple. When you are six years old and two adults you trust the most tell you it's ok to hold snakes, you never ever again have any fear of that animal.

While Rom's work with snakes is fascinating, it is his work with crocodiles that was pioneering. Before Steve Irwin, there was a man called Romulus Whitaker who was a crocodile wrangler.

He is the founder of the Madras Snake Park, the Crocodile Bank in Chennai and the Andaman and Nicobar Environment Trust. One of India's finest conservationists, he was considered a maverick and an iconoclast. With little patience for the stodgy, secretive bureaucratic functioning of the government even within the forest department, he took steps and made decisions that changed the face of reptile conservation in the country.

I was eight years old when I was invited to the crocodile bank to be part of a children's programme on TV on DD, where Rom would interact with us and show us all his crocodiles. Some of us who volunteered were even allowed into the gharial pit with him to feed some fish to them. I will never forget the way my heart thundered in my ears and how sweaty my palms got when those beautiful animals with their long snouts with the bulbous mass crawled towards us, their long bodies undulating gracefully.

Since then, I have been to the crocodile bank several dozen times and I went back after joining NDTV to do a *Born Wild* half-hour piece on crocodiles. As an eight-year-old, I learnt that gharials were very endangered and today, as an adult, I know that the situation is even more grim with them being critically endangered.

Rom had a spectacular vision. He would breed crocodiles in the bank and then release them back into the waters from where they had disappeared. He had one more idea – he would use the surplus crocodiles as meat and skin and have the money made from that pay for the conservation of their wild populations. The second half of his plan was met with great horror and dismay and was never implemented. Ironically, everywhere else in the world that is exactly

what is happening. The crocodile bank today is the Madras Crocodile
Bank Trust and Centre for Herpetology and one of the largest
zoos for reptiles in the world. Its primary aim is scientific research,
education and awareness. The captive breeding of endangered species,
crocodilians and amphibians like fresh water turtles and tortoises lies at
the heart of the trust and they provide captive bred animals to various
forest departments around India for release and also to neighbouring
countries and other zoos worldwide. Their main conservation efforts
centre around the gharial.

In the last seventy years in India, the gharial population has seen a
steady decline with less than an estimated 235 alive today in the wild.
They are endemic to the Indian sub-continent, which means they are
only found here and nowhere else in the world. They are fish-eating
crocodiles, their long thin snouts with the serrated teeth uniquely
adapted to snaring fish. It is the male gharial that has the bulbous mass
on the tip of its snout.

At one time, they were found in most riverine habitats and Indian
river systems, from the Indus river in the West to the Irrawady river in
the East. Today, they are found in less than 2 per cent of their original
home ranges. This small area too is threatened due to how polluted
our rivers have become and all the legal and illegal sand mining,
hydroelectricity projects, water diversion for irrigation and canals.

Many years ago in 2005 I found myself on the Chambal river
near Agra. The Chambal, a 960-kilometre-long river is a rain-fed
river that originates in Madhya Pradesh before winding its way across
Rajasthan and then turning back and heading towards Uttar Pradesh
before reaching the Yamuna. The Chambal is seen as the only relatively
unpolluted river in India. There is an interesting legend behind this.
Its original name is Charmanwati or 'river of skins'. Legend has it
that the Aryan king, Rantideva, sacrificed so many animals during an
agnihotra, that the running blood created a river, the Charmanvati, or
now the Chambal. In the Mahabharat, after the attempted disrobing
of Draupadi, she cursed anyone who would drink the waters of the
Charmanwati river. Thanks to its 'unholy' cultural origins and the
curse, the Chambal is the one river, rather ironically, not desecrated
by worshippers unlike the Ganga. The Chambal has also escaped the
construction of industry on its banks. This has enabled it to become

a haven for aquatic flora and fauna and the Chambal sanctuary is home to the critically endangered gharial, with one of three breeding populations found here as also river turtles and dolphins. However, the Chambal has not escaped human intervention unscathed.

In 2007–2008, over a 100 gharials were found dead in a 70-km stretch of the river. Analysis of the bodies has still not produced any conclusive evidence of why they died. But the deaths have been a blow to the fragile population, tipping the wild gharials closer to extinction. While they do breed well in captivity, and there are several ongoing projects where they are being released into rivers in order to build wild populations, until a viable number of breeding adults are reached, the population is still on the brink.

The mass deaths in 2007–2008 have yet to lead to concrete answers. The necropsies showed that the kidneys and the livers of the dead animals were affected and it is suspected that it might be due to a new unidentified toxin. But analysis of the waters in the area of the river where the animals were found dead did not show any unusual pollutants or change. No other river fauna was affected either. Most of the dead were juveniles and sub-adults and many of them were male. One of the theories afloat is that these animals might have eaten fish that were infected by pathogens and toxins found in the Yamuna, that had swum up into the Chambal river. The gharial is a predominantly fish-eating species and toxic fish could cause massive bio accumulation of toxins which then would get centred in the organs like the liver and kidneys. Analysis of the tissues and organs also showed the presence of protozoan parasites. However such parasites are quite commonly present in these crocodiles and do not usually cause death. The animals also did not look malnourished or ill, leading scientists to believe that the deaths came on them suddenly.

When I visited the Chambal, I saw gharials basking on the sand banks, and the water was a crystal blue-grey; in many places, I could see clear down to the river bed. Terrapins sunned themselves on logs while many birds, like cormorants, snake-necked birds, kingfishers, plovers and black-winged stilts walked, floated, skimmed and flew on the surface. The river lived up to its name of being one of the cleaner rivers of India and the sanctuary teemed with life. This is, however, where the interconnectedness of life starts to play. Even though the

Chambal has remained relatively unpolluted, the Yamuna is a river of toxins. Fish bloated with the river's poison move between the Chambal and the Yamuna, as they have no boundaries, and infect the animals in the sanctuary that eat them. This is one example of why linking rivers to satisfy our need for river transport, or cargo transport or even irrigation and waterways is a bad idea, apart from the sheer magnitude of costs involved and the logistics of linking bodies of water, with whose flood banks and drainage plains we have already tampered.

I got to watch gharial hatchlings come out of their nests and make their way into the river and I wondered then if the young babies would overcome all obstacles in their path to become adults and the future breeding population. But a year later, with the deaths of over 100 gharials in this river, the future of this species seems bleak. People like Rom are fighting for their conservation but in the new India, with the god of development presiding, it is a losing battle. While mass deaths are unsusual, many gharials die every year in our rivers when they get entangled in the fishing nets. In the old days, the fishing nets were made of a fine mesh and the local fishermen knew and understood this crocodile. As the sacred animal of the Ganges, the gharial even enjoyed a certain status of respect and worship. The juggernaut of modern India slowly changed that, though. The fine mesh nets have now given way to nylon nets and the gharial is seen as competition in these waters for the fish. One old fisherman told me that he felt sad that the gharial was not being given the respect it deserved. He said to me that when he was young, he used to feel very happy when he saw the gharials in the water, because he knew then that the fishing would be good. He instinctively understood that the presence of the gharial meant the presence of a healthy ecosystem and therefore, the presence of fish. 'All are god's creatures,' he said, 'the fish, the crocodile, us. We exist together and the river is our world. It has changed and as we lost respect for the river, now it is all cursed.'

Apart from polluted waters, nylon nets and angry fisherfolk, sand mining is one of the main reasons for the crash in gharial populations. While the gharial spends most of its time in the water, it does come out on to the sandbanks to rest and bask. Crocodiles are cold-blooded creatures and need a certain amount of time to bask in the sun everyday to regulate their body temperatures. But more importantly, they use

*A baby seal lying exhausted on a rock after a storm washed him ashore.
We sat with him for a while and then helped him back into the ocean.*
Photo credit: Craig Foster

the sandbanks to build their nest and lay their eggs. The babies, once hatched, find their way into the river. The gharial is an involved daddy. He watches the nest until the eggs hatch and then he guides the baby crocs into the river. He often ferries them on his back and his head. Even mugger crocodile males do the same in India. They even help some of the babies hatch by gently cracking the egg shell around them. They might not stick around for a long time watching their babies grow but they are most certainly there when the youngsters take those first crucial steps from being hatchlings to swimmers. When they become adult females ready to give birth, they come to the sand banks where they were born. Sand mining has shattered this fragile cycle. Sand banks are vanishing and illegal mining has even crushed and destroyed hundreds of nests, putting a full stop on future generations of this beautiful reptile.

Rom, all those years ago, talked to us about the gharial and how the crocodile bank bred them and released hatchlings into rivers where they used to be found. He talked about how this would really help their future, while also talking about how important it was for us not to pollute our rivers and protect the ecosystem. Little did he or I at the time anticipate that the situation would deteriorate to a point where there would be less than 300 breeding animals left in the wild river systems. Nor did he anticipate that there would be resistance to the release of the hatchlings in the rivers.

But if the gharial, a fish-eating crocodilian that has never ever attacked a person, faced resistance, then what of our muggers and the salt water crocodiles?

As a child, it was a thrilling to watch Rom nonchalantly leap into a pit of muggers and hold them off with just a stick. I have watched him handle some of the sub-adults with a deft and sure touch. This was in the days prior to big reality TV. Otherwise, the world would have known and appreciated Rom Whittaker as they do Steve Irwin today as the crocodile man. He is still invited around the world to countries with wild crocodiles and advises them on how to protect them and how to solve crocodile-human conflict.

Today, the crocodile bank holds specimens of all the crocodilian species of the world and has, in fact, a surfeit of muggers and salt water crocodiles with nowhere to go.

My first up and personal glimpse of the salt water crocodile was at the crocodile bank. The salt water crocodile exhibit was impressive and had several dozen animals. The salt water crocodile is the largest of all living crocodile species. They can reach up to the size of 6.5 metres, but on average stay within 4 to 4.5 metres long. Female crocodiles are slightly smaller than the males. While their main habitats are estuaries, swamps and deltas, they can be found in the open ocean. At one time in India, salt water crocodiles were found all the way from Tamil Nadu to West Bengal, but poaching and habitat loss has now restricted its range to just parts of the coast of Orissa in the Bhitarkanika mangrove forests and West Bengal in the Sunderbans. The salt water crocodile is an aggressive predator and in the monsoons, dozens of people are killed when rising water and floods often bring crocodiles literally into peoples' homes.

While they looked fierce and intimidating, it was not until I saw Jaws 3 that I got a glimpse into why the salties are the largest of the crocodilians and are the apex predators of an ecosystem. Jaws 3 arrived at the crocodile bank in the late seventies and was housed with several other crocodiles. He soon outstripped them by growing nearly 16 feet long. Not only was his size intimidating but so was his attitude. He started to attack the crocodiles around him and had to be housed separately. Today, he has an enclosure to himself and still kills the female crocodiles if they are introduced into his enclosure.

The 'Jaws' experience, as I call it, was one of the highlights of my *Born Wild* shooting adventures. Gargi, Siva, my camera person and I climbed into his vast enclosure along with a plump unassuming man casually dressed in a lungi and vest. He asked us to stay back while he sauntered towards the pond. He carried nothing but a metal bucket with meat in it and a wooden stick. At the edge of the pond, he cupped his hand and smacked the water and banged the bucket saying 'Ba, ba... Jaaaaaaaaws' – 'ba' in Tamil means 'come'. While we watched, incredulous, the water started to heave and a mini monster rose from the depths. It was just the head and forebody, and yet the crocodile just dwarfed all of us. The man just casually bashed Jaws on the nose with a piece of meat and proceeded to feed him and then when it was over, walked away. Jaws fell back into the water, but did not submerge completely, keeping a beady eye on us. The man walked to the three of us, who by this time were suitably gobsmacked, and asked me in Tamil what I thought of Jaws. I had no words, except to tell him that I thought he was very brave for just standing there nose to nose with a fierce and aggressive predator. The man just smiled and said he had been doing this for a long time and that he and the crocodile had an understanding.

I could not imagine how that could be possible, because all science teaches us that the reptilian brain is one of the most primitive brains and functions purely on instinct that is governed by hunger, the need to mate and the fight or flight response. Reptiles being biologically cold-blooded and without the parental instincts of a mammal, have always been seen as 'unfeeling', 'cold', not tameable and creatures that depict no emotion. Especially crocodiles are seen as creatures that are always unpredictable and incapable of feeling or ever bonding with a human. They are the only predators that do view humans as prey. I would soon learn that this was not a truism.

Four years ago, my husband Craig and his brother travelled to Costa Rica, to meet a man called Chito. They travelled all the way from Cape Town to Costa Rica because they had heard that Chito had a crocodile whom he had rescued as a juvenile, who now lived with him and shared an extraordinary bond of companionship. Intrigued by this idea, Craig and his bother went to Costa Rica to make a movie about it. What they found would astound them and me when I saw the footage. Chito named his crocodile Pocho. Pocho was a sixteen-foot south American crocodile. These animals have a reputation for being aggressive and have attacked and killed dozens of people. Pocho had been shot by poachers when he was a juvenile and Chito nursed him back from the brink of death. This bond that was created between them lasted for twenty-three years. Chito could swim, play, roll with Pocho in the water and trained him to follow hand signals. In twenty-three years of dealing with Pocho, Chito did not have one scratch on his body from the crocodile. Not once did the animal act threateningly or intimidatingly towards him. Pocho was intimidating and threatening towards anyone else who approached him, but loved Chito in a way that has never been seen before between a man and a crocodile. Sometimes, we humans are so busy seeing animals as species, we forget to understand them as individuals. Pocho clearly had the ability to not just bond but to love.

In Africa, in a village called Paga, in Ghana, the people of the village swim and bathe in a pond that houses more than 1,000 Nile crocodiles. Nile crocodiles are notorious in Africa, responsible for several hundred human attacks and deaths a year. Here in Paga village, however, no one ever gets attacked or killed. It seems like a huge urban myth or legend, seemingly against all laws of nature, but it's true. It is believed everyone in the village has a corresponding crocodile in Bolgatanga, and according to reliable sources, the deaths of important village personalities have coincided with the death of a crocodile. Because they believe crocs are the souls of their village relatives, people never hurt or kill the sacred animals. Legend has it that one of the ancient crocodiles of Bolgatanga saved the first man to settle in the area by guiding him to the pond to quench his thirst after a long journey. In return, the man declared Bolgatanga sacred, and told his people the crocs should be treated like royalty. To this day, it is taboo to hurt or kill any of the reptiles. What's more, it's not just villagers that the crocodiles do not harm but tourists come to Paga in the millions to witness the phenomenon and can sit on the crocodiles and hold their tails and pet them without coming to harm either. Scientists have not been able to understand why this is so in this particular village when everywhere else in Africa, people are killed in dozens by the same species of crocodiles.

As usual, the natural world is throwing us a curve just when we get complacent and think we know everything. Uncle Siddharth used to say to me, if anyone thinks they know the natural world then, well, they know nothing. We only have the privilege of observing, the arrogance of knowing, though, that we don't have. This is a principle by which I have lived my own life when it comes to understanding nature and conservation issues. What seems to be so in one place with the same animal can be entirely different in another.

Around the world, crocodiles like other big predators are finding it hard to survive. Crocodiles are a keystone species, in that their role as apex predators of an ecosystem is key to the survival of that ecosystem. As hatchlings, they are food for dozens of animals and birds and as adults, they regulate the population of various animals in a river ecosystem, keeping them healthy and sustainable. They also burrow, create tunnels, churn up sand, creating various aquatic habitats for the growth of plants and animals. In the dry seasons, crocodiles often know where water sources are or have even created smaller water sources by digging and burrowing. When a major drought hits an area, crocodiles can often go into a state called stasis, where they burrow into the sand or into a cave and stay until the water comes back. For their size, they are not eating machines. Closely related to birds, they have small appetites and can survive on a few hundred grams of meat a week! They live for at least a hundred years if not hunted or killed by other predators and a large crocodile is a testament to survival, cunning and intelligence. River ecosystems in which they are found thrive only with the presence of crocodiles. Removing crocodiles from an ecosystem entirely could lead to the flourishing of other fish and amphibians that will ultimately ruin the system when their numbers are not regulated and often, even the flow of the river can be affected as there are no crocodiles to burrow their way and make room through silt and sand for the water to flow.

It is a horrific thing to be attacked by a crocodile. If the person does not die or manages to escape, they are often left without limbs or with terrible scars. One can most certainly understand the fear and, often, loathing the animals generate. Yet, they are part of this wonderful web of nature and nature has created them to withstand changes and hardships. They evolved 230 million years ago and except for the size of some ancestors, they are virtually unchanged from an evolutionary point of view. Which means, nature felt she had created something perfect and did not need to tamper with the design.

It is easy for those of us who do not live near rivers or ponds with these animals and do not need to go to the rivers and ponds directly for our water, or food or even daily ablutions, to want to save and protect the animals. To get people to live with crocodiles is asking for a lot and the right incentives need to be in place for that. Communities need to be taught as to why the crocodile is important to the river and they need to understand the link between the fish they eat and the presence of the crocodile in the water. They also need to be taught how to best avoid being attacked when they go about their daily work. One of the most successful methods is to not go to the same spot on the river every day. Crocodiles are ambush predators and literally cannot be seen sometimes until they strike. Shallow or deep water makes no difference.

For their safety, of course, it is better for the people to approach clear parts of the river where they can see all the way down to the sand bed, avoid areas where they can identify crocodile drag marks on the sand banks and switch the spots where they bathe or wash or clothes. It is also advisable for them to approach the river in groups and have one or two people armed with heavy sticks to keep a watch and take turns.

Right now the Madhya Pradesh government is toying with the idea of changing the topography of the Chambal ravines and flattening them in order to convert them into cultivable lands for farmers. Apart from the fact that the ravines hold several other species of importance on land like wolves, hyenas, leopards, hares and others, the very topography of the ravines is why the river has the course it does. Much of how the river flows is dependent on the hydrology of the landscape around it. Floods, droughts, rains, all of that reacts to how the land around the river banks stays natural. Many small rivulets that form in the ravines in the monsoon rush into the Chambal, feeding the river as it flows through Madhya Pradesh. Changing this will drastically alter that and also endanger the river by flooding it with silt. With the Chambal really being the last of the wild strongholds for the gharial, this will be a disastrous development.

As human population numbers grow and human development is blanketing the world, the first great casualties are our great predators. We see them as competition, we see them as dangerous, and we see them as the very embodiment of the wild, or that which cannot be tamed, and yet we seem not to see that without them we will have no wild, and by extension, all the benefits of the wild we enjoy starting with clean air, fresh water and food.

Harmony with land is like harmony with a friend; you cannot cherish his right hand and chop off his left. That is to say, you cannot love game and hate predators... The land is one organism.
– Aldo Leopold

And when an animal is not particularly dangerous to us, then, we can rally around to make a tremendous difference. In the 1800s, the southern African coast was home to millions of cape fur seals. By the early 1900s, however, their numbers had dropped to a few thousand due to extensive hunting. Their pelts were prized for the warmth they provided while being waterproof and seal meat was consumed as well. It looked like the cape fur seal was on its way out when protection was enforced from 1893. All hunting of the seals was banned and the populations bounced back. While the existing population of an estimated 2 million seals is still only 10 percent of the original numbers, it is nevertheless one of the conservation success stories. As big as they are and as territorially aggressive as male cape fur seals can be, they have never been an overt threat to people. No one has lost their life to a seal attack. While some people have been bitten on occasion, it has never been fatal. Combined with that, their cuteness has made them poster children for favourite animals to save.

This is a phenomenon in conservation that most certainly translates into which animals get saved and which fall through the cracks. Uncle Siddharth used to talk to me about it all the time. As a child, of course, I too preferred the cute and the cuddly while the creepy crawly and the scaly were not favourites. Through his eyes, I quickly learnt to see that they were all special and that each had a role to play and soon that became what I saw with my eyes as well. You can't love things in nature in isolation, he would say. It is all so interconnected that if you love one thing, you had better understand that the one thing exists because everything else does.

So, while many animals fell through the cracks of conservation in South Africa, the cape fur seal made a full-on comeback. Part of it was the animal's own ability to survive. Once they stopped being slaughtered in the millions, their natural reproductive rates, their ability to be generalist feeders (that is, animals who could pretty much eat most things to survive) and their relative size and speed not making them prey to much but the large sharks, they thrived.

On a beautiful winter day, without a breath of wind in an ocean so calm it looked like a glass surface, I went diving with seals. There is an island fairly close to the shore in Haut Bay in Cape Town, that is surrounded by relatively shallow water, thereby making it safe from sharks. The most daunting thing for me apart from the fact that I barely swim was the water temperature. The lack of swimming skills was made up for by the fact that I was wearing a wet suit that acts as a buoyant device, keeping one afloat, and the snorkel allowed for me to breathe in the water. The water temperature, however, was about 11 degrees celsius. Even through my wet suit I felt like I was encased in ice and any exposed skin like my face immediately went completely numb. I did grow up in the tropics where water temperatures sometimes stayed at 30 degrees! So trying to overcome the idea that icicles were forming in my brain, I slowly started to look around in the water.

The water was crystal clear and it was just teeming with seals. They were everywhere, darting, swooping, flying like some strange underwater acrobatic ballet dancers. They surfed the surface, they leapt through the water and they hurled themselves in every which direction they could. Some of them were curious enough to come and brush me with their flippers as they must have wondered why I was so ungainly in the water! My husband, Craig, who is a water baby, however, received very different behaviour with baby seals actually climbing on and sitting on his back while he dived. The parents watched carefully but never interfered or even appeared threatening because I guess we made no attempt to reach out and touch or grab the animals. We allowed them to come into our space if they so chose.

Fur seals are so named for their thick pelt, unlike true seals which have only a thin covering of hair. Male fur seals are large animals growing to 7 feet and weigh in at almost 300 kilos while the females are much smaller, less than half the size weighing in at about a 100 kilos. There are a few interesting facts about them – they have ears that stick out of their heads; their entire body is a strong supple mass of muscle and sinew that enables them to twist and leap with ease in the water; their flippers are strong, enabling them to swim fast and steer with great agility. On the surface, they often adopt a motion called porpoising – the beautiful, curved, in-and-out motion that allows them to just power through the water – or they dive and then bullet through like torpedoes. They also often lie on their sides with a single flipper rising out of the water like a sail and often just hang upside down (this is when you see just the back flippers stick out on the surface).

And it is when I saw them in the water from such an intimate perspective that I understood why the great white sharks need deeper waters in order to catch them. The great white sharks use the depth to not only perform an ambush attack, but use the depth to propel themselves with great speed towards the surface when attempting to grab the seal. Large bull seals can cause considerable damage to even

A gharial. One of the most endangered crocodilians in the world watching us suspiciously from his spot.
Photo credit: Craig Foster

a great white if they can see the animal coming and feel threatened. A bite from a large bull seal while not fatal can put the shark out of commission and no predator wants to risk injury if they are to stay in prime condition to hunt. Smaller sharks like blue sharks have been killed by large seals that simply rip their gills out. Bull seals, when sexually mature, collect a harem of females and guard their territory quite jealously. They will attack and chase other lesser dominants out of their area when in mating frenzy. It is the females who do most of the parenting of the pups and look after the pups until they are atleast a year old.

Once the females are sexually mature, they have a pup at least once every year. The pups initially are too small to be in the water and are very dependent on their mother's milk for survival. In order to produce the rich milk in sufficient quantities, the female seal has to go out and hunt. The cape fur seal pups do not venture out into the water until they are a couple of months old and have developed the waterproof fur. Females pup between mid November and late December. In the last few years, many pups have been killed by large seas whipped up by storms. Large waves crash onto the islands and wash the pups away.

Young pups who enter the water for the first time are also at danger of being pushed too far into the ocean to get back to their island and get taken out by sharks and often wash up dead on the beaches.

One windy April day, as I finished my walk in Cape Point and decided to rest for a bit on the rocks on the shores of the ocean and watch the waves break, I stumbled on a seal pup on the rocks. Initially, I thought he might be dead but on getting closer, I found he was very much alive, just tired and taking a rest. He was on a high rock quite far from the water and I surmised that a large wave had set him off course and washed him up on to these rocks. After sitting next to him for about an hour and getting him used to my presence, I slowly moved really close and gently reached out to touch him. He made no aggressive moves towards me and accepted my touch. I then gently checked him for injuries and he had none. So, I just sat with him for a couple of hours and when he seemed like he wanted to move, Craig picked him up and we set him in the water and in a minute, he was frolicking and just swam away. It was a stunning moment of connect with the wild and made me appreciate the cape fur seals that much more. Lone adult females and males who do clamber on to the rocks to take a break or a nap are not as approachable and do give you a warning when they feel you are getting too close. But if one is calm and non-threatening in one's approach, you do get the opportunity to sit on the rocks with them and just watch them for a while.

The resurgence of the seal is not without its problems. One would think that when a species comes back from the brink, everyone will be happy and we can all congratulate ourselves on a job well done. But this is not the case and throws up one of the more interesting conservation conundrums, which is, if we have already messed up the web of life, when we endeavour to bring some species back while others are still floundering, are we still in a mess?

The first major problem is the fact that sea birds in South Africa are not doing all that great. The penguins and the bank cormorants are endangered, while the gannets are threatened and other sea birds are vulnerable. These birds are nowhere near their historical population numbers and are also island and select beach dwellers. It is here that they mate, nest and raise their young. In times past, millions of sea birds and seals would have fought for space, been killed, eaten, lost babies, and yet the numbers and the cycle of life would have kept everything going. But the advent of man depleted the numbers of the sea birds, but while seal numbers bounced back, the birds did not. So now the bird enthusiasts want exclusive breeding islands for the birds where they can breed in peace without marauding seals and the seal enthusiasts see it as discrimination. Many islands now have watchers who chase seals off the island and keep them from using the islands used mainly by sea birds as resting or breeding sites. Many sub-adult seals and indeed some adult seals also predate on the sea birds. While seal enthusiasts insist that is only some sub-adult rogue seals who do the actual predation, sea bird conservationists say the threat is quite real. Near Malgas Island, in the waters, cape fur seals regularly patrol the perimeter of the island just as the gannet chicks are ready for their solo flight. They take off from the island and their energy bursts are short and often, they have to land in the open water and wait for wind to assist them with another take off or wait until their strength returns. The seals pick off these young birds. They grab them from below and slap them around on the water. They eat away most of the nutritious fat and the close-to-dead bird is left at the mercy of the ocean, where once they die, they then feed other creatures like the rock lobsters and whelks and others. Off of Dyer Island, cape fur seals feed on both bank cormorants and the penguins. The debate rages on as to whether this predation is actually making a huge impact on the bird populations.

Seals are also now being seen as competition by the fishermen, who lose a small amount of catch to the animals. Seals mainly feed on anchovies, sardines, hake, pilchards and also octopuses, rock lobsters and pretty much anything else they can eat. Seals are being shot by some of the fishermen when they make an attempt to grab the fish and often, several of them get tangled in the nets. Fishermen also blame dropping catch numbers on the seals.

The reality is that it is commercial fishing that is causing most of the problems. By depleting fish stocks they are reducing the food available for sea birds.

But with their population estimated at between 1.5 million and 2 million, Namibia, which has close to 70 per cent of the seal populations of southern Africa, practices seal culling.

South Africa too practiced culling until public pressure ensured the practice was banned in 1990. Despite international condemnation, Namibia continues to cull seals. Close to 85,000 pups and six thousand adult bull seals are slaughtered every year. The pups for the soft luxuriant fur and the bulls for their penises that are valued as an aphrodisiac in the Far East. The meat is also consumed.

While I have not witnessed an actual cull, I did meet with filmmaker Craig Matthews a journalist and filmmaker, who exposed the culling in Namibia when he risked his life to film it secretly.

The footage is chilling. A group of men herd the seals and the pups and females run in panic. They are forced to run in one line and in one direction. As they run, they are clubbed over the head with a big stick and then stabbed with a knife. This is called the stun and stick method. It is seen as a humane way of killing the animal. But what I saw was nowhere close to humane. The seals had to be clubbed several times and hard. Then they were stabbed repeatedly, the pups scream for their mother and then scream in pain, while the females are distraught from the stress. The big bull seals, owing to their size, are simply shot. The reason they will not shoot the pups is because bullet holes will mar their beautiful pelt. It was an eerie watch, the Namibian beach wreathed in mist. Men dressed in blue overalls, carrying heavy sticks and knives and a heaving terrified mass of seals.

The cull quotas are set at 90 per cent of the born pups, and every year it ranges between 80,000 and 90,000 pups and 6,000 bulls. In the last decade, seals have also been dying from starvation as Namibia's over-fished coast is just not providing them with enough food. So many conservationists question the population numbers posted and have been fighting the cull, which they call a harvest.

Namibia makes no explanation as to how the quota numbers are arrived at. There is annual census, carried out from the air and a population estimate is made. From the numbers set in the quotas, it is to be believed that a hundred thousand seal pups are born every year.

The harvest takes place in Cape Cross and Wolf/At Last bay where 75 per cent of the pups are born. The harvest takes place between the months of July and November, when the seal pups are between six and nine months old.

*Illegal and extensive sand mining of river banks is one of the main reasons for
loss of habitat for animals that nest in the sand banks and is changing the
course of our rivers, exacerbating flood events.*
Photo credit: Kalyan Verma

While seals are protected from hunting and commercial harvesting
in South Africa, it does not mean that they are safe in the waters around
South Africa. 65,000 trawl nets are deployed in and around South
African waters. Often fishermen will follow foraging seals to locate
the shoals of pelagic fish they need to catch and then simply throw in
the trawl net, trapping the seals with the fish, and sometimes up to a
100 seals are caught in the nets. The ones that do not die of drowning
are either shot or beaten to death and many get injured. The South
African Commission on Sealing found that at least 30,000 seals are
drowned by just one sector in a thirteen-sector fishing industry.

Apart from trawl nets, discarded fishing lines, nylon ropes, hooks,
plastic trash are entangling and injuring and killing seals annually. I
have seen so many seals with deep gashes around their necks because
nylon lines have wrapped around them getting tighter and tighter,
cutting into the fat. It is hard to get these lines off of them as they are
ocean-dwelling mammals and are not easy to get close to.

Added to the cull numbers and the starvation deaths,
conservationists question whether present population estimates of 1.5
million to 2 million animals are accurate. It is, however, a fact that
population numbers have stayed stagnant at these estimates for over a
decade now.

As of now, their IUCN status is of least concern as population numbers are still seen as healthy.

Seals are one of the big oceanic predators and their presence helps maintain an ecological balance in the oceans. The notion that increased seal populations are responsible for loss of fish stocks is patent nonsense. The Namibian cull, apart from being a harvest for commercial purposes, is still carried out as the government insists it is to protect fish stocks from increasing seal numbers. A few scientific studies done on seal culls have proven this theory entirely false. Marine ecosystems are complex, with seals and other top predators playing important roles in maintaining an ecological balance. The studies that have been done show that predatory fish are the greatest predators of other fish. A study of a grey seal cull in Iceland from 1990 to 2002, in which 66 per cent of the seal population was killed, showed that there was no effect on the cod stocks that the cull was intended to increase.

The plain truth is that we humans are the biggest reason for plummeting fish stocks. Since 1950, one in four of the world's fisheries has collapsed due to overfishing and 77 per cent of the world's marine fish stocks are fully exploited, over exploited, depleted or slowly recovering.

Scientists estimate that 90 per cent of the world's large fish have been removed from our oceans, including many tuna, sharks, halibut, grouper and other top level predators which help maintain an ecological balance.

Of the 3.5 million fishing vessels worldwide, only 1.7 per cent are classified as large-scale, industrial vessels, yet these vessels take almost 60 per cent of the global fish catch. Industrial fishing fleets kill and discard about 27 million tons of fish on average each year. That means that one quarter of the annual marine fish catch is thrown overboard dead. If that figure is hard to imagine, let me put it to you like this, that it is like throwing the equivalent of three million African elephants overboard from a ship dead. Every time we order shrimp, we must be told that for every kilo of shrimp landed, over 10 kilos of tropical marine life is caught and dies.

It's a mess of oceanic proportions and we are the reason for it, not the seals, not the sharks, or any of the other magnificent oceanic predators, or predators in the river ecosystems like crocodiles are to blame.

And for anyone who feels that the ocean is beyond their comprehension or that they don't really have much to do with what is happening in it, well, no matter who you are, if you live you need oxygen to breathe. No matter where you live, even if it is in some landlocked place far away from the ocean, a large amount of the oxygen we need comes from it. So no matter where you are, the ocean touches you and you cannot escape that one single fact.

And it is an interesting biological fact that all of us have in our veins the exact same percentage of salt in our blood that exists in the ocean and, therefore, we have salt in our blood, in our sweat, in our tears. We are tied to the ocean. And when we go back to the sea – whether it is to sail or to watch it – we are going back from whence we came. [Remarks at the dinner for America's Cup Crews, 14 September, 1962]

– John F. Kennedy

9

COLD DESERT, HOT SALT PANS AND THE KAROO

NO LIFE WITHOUT A LANDSCAPE

I think the kind of landscape that you grew up in, it lives with you. I don't think it's true of people who've grown up in cities so much; you may love a building, but I don't think that you can love it in the way that you love a tree or a river or the colour of the earth; it's a different kind of love.

– Arundhati Roy

WHEN SOMEONE says it cuts like a knife, I am sure they are referring to the wind that howls down the Changtang plateau. Both desolate and stunning, this expanse of cold desert in Ladakh is one of my favourite places on earth. It is a unique ecosystem because while it is a cold desert, this area contains high altitude wetlands. While the land around is flat, the towering peaks around me are anything but. Ice covers the upper reaches, lichen and small green plants burst out in clumps, and the Tso Kar lake reflects back the empty infinite sky in a bright turquoise. I feel like the last human on the planet, waiting for the land to just rear up and swallow me.

When I was fifteen years old, I ventured into the Himalayas for the first time. I had on previous occasions visited hill stations in the summer on holiday, but this was the first time that I was going on a Himalayan trek and would be in the heart of the mountains that were the backdrop to all my various holiday destinations. I had two reasons for going on this trip. One was that my school, Rishi Valley, had organised it and I knew that it would be a great opportunity for me to see and experience the mountains in a way I never had before. The other reason was very personal. A few years before my Himalayan trek, uncle Siddharth had been diagnosed with cancer. I watched him fight it with the same indomitable will and wild spirit he showed when chasing his favourite birds, the brahmini and pariah kites. It looked for the longest time like he had it beat and we hoped that he had entered remission, but then it came back.

That summer, just before the start of my ten-day trek, things had been really bad and he was mostly in hospital and, this time, it looked like he had no fight left in him. For me, it was the most horrible summer of my life. There were no more walks, no more recording of frog sounds in the rain, no more stories of animals and the wild, but large empty silences, a void that nothing could fill. On one of his more energetic and lucid days, he asked me if I had a plan for my holiday. I said I was thinking about going on the Himalayan trek organised by school. For the first time in a long time, he perked up and said, 'Swati, the Himalayas are special, maybe it's because they are very young as mountains go, that they have an energy and a quality that you will find on no other mountain. Get me a walking stick and a Kulu cap and some apple jam.'

A few days later, when I went to see him again, just a day before he would go back to the hospital, and two days before my trek, and he said to me, 'Respect the mountain, it is in human nature to feel like one has achieved something or conquered the mountain by climbing it, but just remember, if you reach the top and complete your trek well, it is because the mountain has allowed you to do so. Don't just walk, young lady,' he said, 'absorb, enjoy, and know that these mountains as young as they are have stood for millions of years and will continue to do so when you and I are dust in the wind.' As I was leaving, he said, 'You do know that they are still growing right? Just like you.' My last memory of him was from the hospital bed, whispering '...trekking means walking, so use your feet' and then he held up his thumb, I knew what he meant, walking and not hitching... and I laughed. I did not know that that would be the last time I would ever see him. The entire trek he was with me and even now when I am in the mountains, he is with me.

Standing on the edge of the Changtang plateau on the banks of Tso Kar lake with brahmini ducks, a pair of black-necked cranes, I could feel him in that wind. Nature is not for the faint of heart, it is for those who want their heart filled, shaken and changed, he was saying.

This cold desert landscape is by far the most stunning landscape I have ever been in. Although I was already at an altitude of about 16,000 feet, while standing on the shores of this lake, mountains that rose easily another 2 to 3 thousand feet towered over me. The mountains were so tall that I could barely make out their tops, wreathed as they were in clouds, and with the shifting clouds and sun, they changed colours like chameleons, from green to red to blue in a magical display.

Lake Tso Kar is one of three lakes in this region. The other two being Lake Tsomorir and Pangong. The bulk of the Changtang plateau extends into Tibet and only a part of it lies in India.

The lakes in Ladakh are the highest salt water lakes in the world. The salt comes from the chemicals found naturally in the soil and the rocks. The fact that the lake exists essentially in a cold high altitude desert system means the lake experiences high levels of evaporation, making the salt content more potent. It is only in the short summer months that the glacial melt adds fresh water to the lake and rain water streams exist for an even shorter period. No fish species have been detected in the lake as the extreme fluctuations of temperature and evaporation changing salt content make it mostly inhospitable. There is, however, much phytoplankton and zoo plankton in the water, making it a feeding ground for various migratory birds and the breeding ground for the black-necked cranes that are only found in India and Tibet. Some 225 species of birds are found in Ladakh, many of them venturing into the high altitude wetlands. High in the sky, lammergeiers and golden eagles circle looking for food. Golden marmots and many other small voles and mice hide in deep burrows in the ground.

The marshes and the pasture land that open up during the summer months attract the Kyang or the Tibetan wild ass to the shores of the lake. We were very lucky to witness a herd of asses come down towards the lake to graze. These wild asses look very similar to the wild asses I have seen in the little Rann of Kutch. While the land and mountains look bare and can be brutal in the winter months, they are still home to some of the most spectacular wildlife in the world. The mountain slopes are home to the snow leopard, the Tibetan wolf, the Eurasian lynx and brown bear. All of them endangered. The slopes with their sparse vegetation are also home to the baral, the blue sheep, which are most commonly seen, the urial, a rare goat found in the lower elevation and so easier prey for hunters and also seen as compensation for the domestic livestock. The ibex, found in the higher craggier mountains, the Tibetan argali sheep and the chiru, the Tibetan antelope, also critically endangered as it is slaughtered for its fine wool that is used to make the shatoosh shawls that are highly prized in north India.

Dr George Shaller, wildlife expert extraordinare, was the first person to make the link between the chiru and the shatoosh trade and today, it is illegal to hunt the chiru for its hair. Extensive media campaigns have also tried to bring home the message to the consumer that shatoosh is illegal and has tried to make the consumer understand the true price paid for the shawl. Shawls that have been in the family over generations prior to the new law were allowed to be declared to the government and stamped as old before the establishment of the law. This has helped reduce the slaughter of the chiru, but they are still critically endangered and poaching is of big concern. The rest of the ungulates in the mountains are still hunted and the advent of roads and highways to some of the more inaccessible parts of the mountains has made it easier for illegal hunting parties to get to these animals. While the population numbers in India are still under protection and there is some control, the larger populations spread across Pakistan and Afghanistan are in great danger.

The increase of human activity in the mountains is also threatening the fragile ecosystem. Earlier, it used to be impossible for people to get to most of these areas. Now tourists are everywhere. There are hardly any checks in place and the mounting garbage generated by tourists is a huge problem. Jeeps are also driven all over the plateau destroying small nests, marmot burrows and young budding flora as most people have no idea of what they are doing. The roads and highways were initially built out of necessity as the area is of great strategic importance – it borders with China and the army needed to have a base there. Not meant for large numbers of people, this region is now buckling under pressure.

On the Changtang, where we were, the Rupshu Plateau, the area is home to a nomadic people called the Changpas. The Changpa are semi-nomadic pasturers. They raise herds of sheep, yaks and goats and horses. The Changpa used to collect salt from the lake and trade it in Tibet. They also used to move freely across the Changthang, prior to the Chinese takeover of Tibet, when they closed the border. With the roads and the army and Leh and Kargil becoming tourism destinations, salt is now found freely in the valley, removing the one big barter item the Changpa used to trade for food. In the harsh winters, they live in their yak tents, and make the tent weather-proof by rubbing yak butter all over it. The tents are so heavy that when they move, it has to be carried in different sections. Many Changpa now live in permanent homes and raise Pashmina goats instead.

As I was standing on the banks of the lake, I heard a strange sound, soft, floating just below the howl of the wind. I looked up into the mountains around me and I saw small figures moving down the steep slopes. First it was some yak, then it was some goat and sheep, and then some men on horseback, herding the lot. The goats, sheep and yak made a beeline for the soft pasture grass around the lakes and the men stood or sat on their horses watching them. The Changpa language is a dialect of Tibetan and our army guide found one man who could speak a little bit of Hindi from his days as a salt trader. He told me how times had changed and that collecting salt was no longer their main work. They now traded the pashmina wool that is used to make pashmina shawls. The pashmina goats, however, required the soft grass for feeding and now unfortunately the kyang herds that also come down to the lake to feed on the grass are being seen as competition. The Changpa insisted that protection had increased the size of the wild ass herds and felt like their goats were not getting enough food to feed on. We had heard disturbing rumours about kiang being shot, but the Changpa I spoke to of course denied that they killed the wild asses. The Tibetan wolves and the lynx and the snow leopard were also in danger as they do tend to predate on the goats and sheep.

The kiang are the largest of the wild asses and are endemic to just this part of Ladakh and parts of Tibet close to the border. They are related to the wild asses found on the Rann of Kutch. At one time, it was thought that they might be a sub species but greater scientific studies on a molecular level have found that they are a separate species. It was unbelievable how I could be standing on two starkly similar yet entirely dissimilar landscapes, looking at two species of wild asses. The one landscape, a massive salt pan, flat and forbidding, boiling up to 50 degrees celsius in the summer months and the other, a cold high-altitude plateau, equally flat but rimmed by mountains taller than it. The one set of asses withstand very high summer temperatures and

Wild asses in the Rann. A mother and her foal.
Photo credit: Kalyan Verma

survive on the Rann and the other withstand ridiculously low winter temperatures and survive on a snow- and ice-blasted landscape.

Standing on the flat Ladakh landscape, watching the wild asses run, I was transported back to the the Rann of Kutch. Equally flat but with no mountains to break its bleak stretch to the horizon. It was in the middle of summer and I was on a camel with Gargi and both of us thought we were seeing things when through the shimmering mists of hot air, peculiar looking creatures walked towards us, paper thin bodies with big bobbling heads. Of course, it was an optical illusion through the heat waves and as they came closer, we made out the robust bodies of the wild asses.

Wild asses look nothing like horses; rather, they look more like zebras in their shape. They have white bellies and light brown top coats. The kiang seemed to have darker brown coats almost gleaming chestnut while the Rann asses seemed to have a more caramel gold coat. The Rann's asses do not have a specific predator as there are no cats big enough to bring them down once they become adults. Their babies can sometimes fall prey to caracal but it is rare. Legend has it that these lands were bare of the asses until Babur the Mughal emperor marched into India and brought them with him. Today, however, several hundred years later, here they are and adapted to this landscape. Packs of feral dogs from nearby human settlements are their biggest danger. There might have been packs of wolves at one time on the Rann, but they have been mostly wiped out for a long time.

The Rann has its own challenges, just like Ladakh. Here, salt workers converge in the summer to extract the salt from the land. Over 50,000 people at a time work on the land and there are trucks going in and out to carry the salt. All of this a rather shocking burden on a very fragile landscape. This co-existence has been going on, though, for centuries now, and asses and people seem to have come to an understanding. The Agariya community whose only means of livelihood is the salt extraction work, meet about 75 per cent of India's salt needs from here. The Rann is the largest salt pan in the world. Speaking to a salt worker here, I learnt that they work back-breaking hours, over fifteen hours a day sometimes. The ground water in the Rann is almost ten times saltier than the sea water and is pumped out using bores. The water is then run through small fields about 20×25 metres. Here the water evaporates slowly under the relentless sun and a silvery substance is left behind. This is the salt which is then collected by the Agariyas. This salt is collected and sent to salt factories and chemical factories across the country by trucks and trains; each Agariya family might take care of 60 such fields. But although without them no salt would ever make it to the markets, the large supply does not ensure a good livelihood for the Agariyas. Most families make a meagre 60 rupees per tonne of salt.

Watching them, women and men in that 50 degrees heat, bend and scrape and stand in that rather acidic water, it made me sad that this hard work could not do more for them and their families. To me this is what's really wrong with how things pan out not just in India but across the world. If the local people cannot be happy and have a decent standard of life, how can we possibly expect them to be the champions of the wildlife and the wild landscapes they live near? When do they have the time? Or even the inclination and the energy? And the Rann does not offer even this to them all year round. In the monsoon, which can last upto four months, the entire Rann becomes a quagmire. Just fifteen metres above sea level, in the rains, it is covered with standing water. The sweet fresh water of the rains which feeds the rivers that flow towards the Gulf of Kutch mix with the salty sea water coming in from the Gulf. This combination of brackish water and sweet water gives rise to the birth of hundreds of millions of shrimp. The same landscape looks like a vast pond filled with migratory birds. At this time of the year, several fisherfolk venture into the Rann to harvest the shrimp. Of course, without the monsoon and this water, the ground water of the Rann would not get replenished and then there would be no more salt to be extracted, but for the Agariyas, the lack of work in this season is a blow and they are one of the poorest communities in India. A clear case where sheer hard work does not lead to a better life.

The rains, however, create another miracle. The Rann is the world's biggest breeding site for the beautiful pink flamingos that flock here every year. The shrimp they consume makes their plumage an even darker pink. The greater flamingos come here in the thousands between August and October and finally leave in March with their chicks. This colony in the Kutch is so large that it is actually called the 'flamingo city'. It is a sea of pink everywhere one looks and the birds, like graceful ballerinas, delicately step back and forth while using their huge beaks like filter feeders to suck in the shrimp and other organisms in the water. It is one of the most beautiful wilderness events in India and now it faces a huge threat.

In 2011, experts and the then wildlife board rejected a proposal to build a road through this part of Kutch. It was unanimously declared that such a road would be ruinous to the fragile area destroying a rich wild landscape. In 2015, however, a go-ahead nod was given by the present government and Wildlife Board. The road will damage the flamingo breeding site, spelling doom for the flamingo population as this is their largest breeding site. This exquisite unique world houses not just the flamingo but the desert fox, the great Indian bustard, the caracal, last remnants of the Indian wolf, some hyenas and the desert cat. All endangered and struggling to survive. It also houses the shravan kavadia, a type of mangrove found nowhere else in the world. This is the only land growing mangrove and in 2013 it was identified as a potential biodiversity site. The very fact that the flamingos come here and breed points to the fact that the influx of sweet water brought in by the rains is crucial for the mix with the salty water from the Gulf. A road could seriously affect the way the water flows and drains. This, in turn, will affect the fragile fresh water-salt water balance affecting the entire food chain, upsetting the whole order of the system.

The government has said the road is necessary as it will be for the use of the border security forces, crucial for the defence of the nation. Experts have pointed out, though, that there is an alternative route that can be used that will have far less effect on the flamingo city and have also pointed out that there is an existing route already in use by the security forces.

While security and the defence of the nation are very worthy causes and yes, the area is only 30 to 40 km from the Pakistan border, the fact that there can be an alternative must most certainly be looked at. The loss on the other end is incalculable otherwise.

The Great Rann of Kutch, along with the Little Rann of Kutch and the Banni grasslands on its southern edge, is situated in the district of Kutch and comprises some 30,000 square kilometres (10,000 square miles) between the Gulf of Kutch and the mouth of the Indus River in southern Pakistan.

Which is why roads into this part of Ladakh have also caused havoc. The kiang, like the Rann wild asses, prefer the open flat plateau area as opposed to the steep mountainous slopes. This might be because it is easier for them to run and navigate the flat lands unlike the sure-footed barals and urials and the other ungulates on the slopes. Their main predator is the Tibetan wolf and their defence is their ability to run fast for long distances and to be able to kick with their hind legs. Wild ass herds are usually divided with the sub-adult males forming a group of bachelors and the females and foals and young males forming other herds. Male stallions that are sexually mature usually are solitary and have their own territories they guard from other males. The male stallion has the sole right of mating with the females in the territory he claims. He will fight or run off any rival. Females usually have one foal each. The status of the wild asses in general has been listed as endangered, with very little known of the populations in China and parts of Tibet and Mongolia.

The older Changpa said that they had ways of living with the wild animals but that things were changing. They saw the roads as a major problem as their young men were leaving the Changpa lifestyle and moving away. The trade in pashmina wool was also a new venture and, being quite lucrative, it was making them more settled and less nomadic. It was also eroding their traditional tolerance of the predators and other wild ungulates. Most of the Changthang today is also a wildlife reserve, protected under the law. The area under protection is larger than Spain and this has helped increase the numbers of the endangered species found here, but as access increases to these areas, human pressure and poaching has taken its toll.

Standing in the lee of the bitter wind, I have to say that at least I understood where the Changpa were coming from. Not a large group, there are less than 6,000 Changpa in the area. In their defence, I at least understood that most of their life was lived as close to the land as possible and the occasional shooting or killing of their wild co-denizens of the land was more survival than commercial or pleasure. The poaching by outside parties, mainly geared to the Chinese trade markets, was more worrying.

The only saving grace was that even now with the all the threats, the habitat is vast and still viable, enabling the growth of both predator and prey population. However, if the out-of-control tourism and lack of rules and general protection is not stepped up in the area, this habitat too is going to suffer.

It was actually amazing to consider that I could be standing at over 5,000 metres in an almost barren cold desert landscape, feeling like the last human on the planet and knowing that the reality was so vastly the opposite with more and more human activity taking over these places and ruining something unique and fragile.

It is key to understanding conservation that the loss of the actual species, while a blow, is not the irreversible loss that the loss of habitat is. Of course, extinction is irreversible in some ways, but an attempted breeding and rewilding can be done as long as habitat stays intact. Take out the habitat and nothing can be done about anything.

Today, in England, they are contemplating rewilding the country side with lynx, 1,300 years after they disappeared from the landscape. The idea is still in its nascent stages but how fantastic if it happens. However, the sticking point is the availability of habitat, vast and wild enough that the lynx don't target the sheep, calves and goats now on that landscape that belong to humans. Very quickly, the rewilded lynx will vanish again.

If there is a place where rewilding is a daily phenomenon, that would be Africa.

The first time I stood under a dome like sky on the vast expanse of the Karoo in South Africa, I felt like I could just blow away like dust and merge into the landscape. There is something so wild and evocative about it. It is both desolate and welcoming. It is barren and filled with life. The Karoo is a semi-desert natural region found only in South Africa. It spans 40,000 sq kilometres. Just like the Rann and the cold desert, the silence is absolute. The Karoo is the one place here in South Africa where I feel the same sense of sweeping glory of a landscape as I do on the Rann and in Ladakh. It's a landscape of craggy mountains, hillocks and an ancient seabed. At night the sky feels like it has dropped down on your shoulder like a diamond encrusted mantle that you trail as you walk through the land. Maybe, I feel such a pull to these landscapes because in their wildness they show us the geological forces that shaped our world and both India and Africa were a landmass called Gondwana. All three of these places have evidence of the ocean having flowed through at one time. Deep in the Karoo or high up in the Himalayas, the ocean once covered it all. The Rann still welcomes the ocean every year during the monsoons.

The vegetation is xerophytic or vegetation that has adapted to survive and adapt to water scarcity. This vegetation is a low yield vegetation and canot sustain a large biomass. Typically in the Karoo animals would wander through mostly in the short rainfall periods in large numbers and the year round herds were small in numbers. The Swartberg mountain range divides it into the Little Karoo and the Great Karoo.

Both the Great and the Little Karoo lie almost entirely within two of the eight South African biomes, the Succulent Karoo Biome and the Nama Karoo Biome. The Succulent Karoo has the world's richest flora of succulent plants. Succulent plants are plants that are fat or fleshy either in their entirety or in certain parts. They are built like this in order to retain water in very arid ecosystems. The Karoo has about one-third of the world's approximately 10,000 succulent species. The region is also extraordinarily rich in plants with storage organs or geophytes. These are plants that are specifically modified to store energy usually in the form of carbohydrates. Geophytes are often found underground in order to protect themselves from herbivores. The Karoo vegetation is classified as an endangered ecosystem.

When the first white settlers arrived in South Africa and moved into these areas, they realized that the succulents made excellent grazing plants for their sheep. Most of the Karoo gradually started to become sheep farms losing its unique vegetation. The settlers also hunted the herbivores for meat and shot the carnivores in order to protect their sheep. The habitat suffered some irreparable losses for decades. Recently, in the last few decades, the sheep farms have been dying out and much of the land has been bought over by private people who have allowed the regeneration of the original vegetation. Slowly many of the farms were joined together to make large conservation conservancies and all the fences between the boundary lines were torn down.

Once the vegetation was allowed to come back, the private owners started to re-introduce animals that were once found here.

I was in one such conservancy called Sanbona. Sanbona is 50,000 hecatres of wilderness at the foot of the majestic Swartberg mountains. Standing as I was on a dry river bed with tumbled quartz stones in the shadow of the stunning Swartberg mountains, I was a mass of shivers. It was deep in winter and a huge cold front had moved in. The wind was icy and cold rain was trying to find its way into the collar of my jacket. While all the external discomforts were certainly hard, I was almost unaware of them as I was standing about 2 metres from a mother cheetah with three cubs. The cubs were sub-adults and had just feasted on a springbok kill. The mother was radio collared. She was a first generation wild cheetah in Sanbona, transferred from another park. She had not only adapted to her new surroundings, she had cubs. The success of any rewilding programme is not only in the successful translocation of the animal but in how they adapted and then went on to breed. The cheetahs were wary but not particularly upset at my presence. I was there with an experienced ranger and followed all his instructions on how to act around them. No loud voices, no sudden movements and definitely no running if they suddenly came close. A cheetah might look like a delicate cat but they are super strong and formidable. They are however less aggressive than the other big cats. Having eaten, the cubs were frisky and wanted to play. The cold weather was also making them hyper active. In India we lost them in the wild decades ago with the last Asiatic cheetah being shot in 1957. For a long time they were bred almost like hunting dogs and used to run down prey for hunting. Babar the Mughal emperor is rumoured to have bred over 400 of them. It made me sad that our empty grasslands in India would only ever feel the ghostly breath of this cat in the wind without the majesty of the fastest land animal that has ever lived on earth. I stayed with the mother and cubs for almost two hours. They didn't run away from us and almost walked in parallel maintaining that 2 metre distance. As we left the cheetahs and walked over a hill we started to see tracks and signs of white rhinos. So we walked very slowly and quietly until we came upon a small group of them grazing. These animals too had been introduced here. The whole of Sanbona now has lions, cheetahs, leopards, rhinos, elephants, giraffes, buffalo, springbok, gemsbok, eland, hartebeest and so many others. All of them reintroduced after the habitat was allowed to regenerate. It was mind boggling to be standing in a wild place that had lost all of its wildness and was empty of everything that made it special less then thirty years ago. It had been ruined and domesticated by humans and now it was wild again and free also because of humans. Sanbona is by no means the only such rewilded place. Many parts of both the Little and the Greater Karoo that had been made into farms have been reconverted into wild places.

The original people of the Karoo were the San and Koi San or the bushmen. Hunter gatherers who lived on this land in harmony with the ecosystem around them until they were killed and persecuted by the new settlers. Even today there are caves in these mountains with the rock art left behind by men and women who moved with the rhythm of the wild.

Today the Karoo faces a new threat, fracking. Fracking involves digging down into the earth before inserting water at a high pressure in order to force open existing fissures to extract natural gas. It is a very destructive process leaving the land in shambles when its over. The South African government has announced that fracking for shale will begin in 2017 in the Karoo. The Karoo is believed to hold 485 trillion cubic feet of shale gas. The drilling has been delayed for both environmental and economic concerns. The South African government has stated that shale gas extraction and usage is the answer to the power crisis that has plagued the country for the last few years with huge power cuts becoming the norm.

Environmentalists say that the move is a disaster as the extraction process has the potential to posion the Karoo's existing ground water.

The Anglo Dutch oil and gas company Shell has said it will invest 200 million dollars in the first step of exploration. With that kind of money involved it seems like all other concerns might get swept under the proverbial rug.

The Treasure Karoo Action Group has brought out a startling statistic claiming that there is no good way to go about extracting shale gas and that it stands to ruin the environment irreversibly. They claim that one well involves the use of 20 million litres of water for both the extraction and the cooling down of the powerful drills. 2,500 trucks is the estimated number required. One pad is set to have 32 wells and ten pads are mooted in one development. They say this will affect 52 per cent of the Karoo.

Fracking is an extremely water intensive process and it is a fact that the Karoo is one of the most arid landscapes on the planet. The little ground water is a precious collection of thousands of years of rain. It also seems not equitable when the landscape will be ruined forever for a resource that will eventually run out. The habitat lost can never be regained and then never rewilded once fracking ruins it.

I have been to Ladakh three times in the last fifteen years and to the Rann twice. Each time, the situation looks worse. There are just more and more people everywhere and the wildlife is receding. There is not a single spot now near the wonderful lakes where one will not find some garbage, most especially, plastic wrappers left behind by some chip-munching Jane or Johny. The accessibility due to construction of roads has also brought in the tourists in droves and with no real regulations or even monitoring in place, tourism is doing more damage in these fragile ecosystems than being a boon to this area.

Tourism is turning into a double-edged sword. On the one hand, the homestays and organised trekking trips really help boost the local economy, but on the other, the unregulated side of it is bringing nothing but pollution and headaches to the locals. Drunk men in four-wheel drives careening around on the Changtang plateau is a huge disturbance to the wildlife and also the fragile ecosystem. The accessibility is also leading to the illegal hunting of the urials and bharal.

The apex predator here in Ladakh is the snow leopard. Rare, shy, almost impossible to spot, this cat is the king of the mountains and is disappearing even before we can understand enough about them to save them.

The snow leopard, the wild asses, the flamingos, and the wild Karoo I wonder if they will all soon just be remnants of a vast and wonderful world that will fade into the mists of history and become legend. The tragedy of that cannot be put into words by me.

Take a quiet walk with mother nature. It will nurture your mind, body and soul.

– Anthony Douglas Williams

THE PERMANENCE OF IMPERMANENCE
MY MOST IMPORTANT LIFE LESSON

Man is the most insane species. He worships an invisible god and destroys a visible nature. Unaware that this nature he is destroying is this God he is worshipping.

— Hubert Reeves

SAND WHIPPED around me, buckets of rain, sheets of water in cascading waterfalls pounded down on my head, sluicing down over

my body, the wind literally knocked me back every time I tried to take a step. The ocean was a grey churning mass and there were puddles on the beach, the sand simply unable to suck up anymore water. I was standing on my local beach in Chennai, with my father and uncle Siddharth in the middle of a raging cyclone. Why those two crazy people decided to walk to the beach in the teeth of a cyclone and drag me along was a mystery. Being thirteen, sulky and irritated, I could not believe that this is how I was going to spend a few hours of my weekend, being lashed by the elements on my local beach.

But something happened while walking the short distance to the beach, something magical. The sulky teenager became ensnared in the complete wild drama of nature and what seemed like extreme torture suddenly morphed into extreme adventure. Every step forward was a battle, every breath misted with water. The main fury of the storm had come and gone at night but the tail end was still spectacular. I could see drenched birds huddled in the trees and stray dogs trying to find shelter under steps and bushes. On the beach, it was just a wild fierce wind with stinging rain.

'This is the voice of god,' my father said, over the yell of the wind, 'this is the touch of god, when sand and sea spray mixed with the rain hit our faces, and these are the tears… the tears,' he said, wiping the rain off his face. Uncle Siddharth laughed, 'It's nature at her best and most potent. Along with all those sunrises and sunsets, the absolute miracle of the everyday ordinary we take for granted, we need a slap on the head from time to time and here it is.' We stood and watched the roiling sea and I sincerely hoped that none of the fisher folk who lived in the village just up the beach had ventured out and been caught in this fury.

It was just the three of us on the beach and not another soul. It felt like we were the last people on a planet that was falling apart, torn asunder by the forces that shaped it. The planet came into being violently, and when it finally dies, it will go just as violently. It is arrogant of us to imagine that we are killing this planet. With our lifestyles what we are doing is actually only killing our ability to survive on this planet. 'Nature will evolve beyond us,' said uncle Siddharth, 'she will always survive because she learnt to adapt and grow while we are still learning how to live. I brought you here today, in this storm, so that you can stand face to face with god and understand that all our actions, while damaging to nature, do not mean her end, but only ours.'

If there was one thing I learnt that day from him it was about the permanence of impermanence. The very essence of what nature is. Not something that was trapped by the rigidity of living but a being who exploded free with the adaptability of survival. School taught me many things but if I look at my life today, I can say with great confidence that my most important lessons were learnt in nature.

Years later as an adult, I was face to face with the permanence of impermanence on the flood plains of the Savuti marshlands in Botswana in Africa. It was the tail end of winter and it was still months before rain would come to these marshes. Thirty years ago, the Savuti Channel dried up and all the water to this area of the Chobe National Park in Botswana stopped. Slowly, drop by drop, water evaporated, leaving this area parched and bereft. Thousands of years ago, this entire area was one massive lake which in itself had vanished with a few tectonic shifts of the earth's plates. Post that, the area slowly became a vast marshland with a channel that flowed through it watering the plains. This water came all the way from Angola, about 1500 km away. If it rained well in the Angolan highlands, then water flowed to the Savuti Channel. It had rained well for years and then it just stopped. All the animals that used to flock here on their way from Moremi, in the delta of the Okavango on their way to the Chobe river and back, had to learn to adapt through the stretch when it suddenly lost water. The animal numbers dropped drastically and many started to avoid this route.

The apex predators of this land are the lions and they too had to adapt to this fresh change or die. They learnt to do something extraordinary, something that happens nowhere else on earth. Here the lions learnt to hunt elephants. With the drop in the numbers of antelopes, with other predators like hyenas, leopards and wild dogs also competing for the scarce meat, the lions had to evolve. Leopards can survive off prey as small as small birds, voles, moles, mice and others, and the occasional buck that wandered by. Being solitary, she only needs to hunt for herself. The wild dogs who have some of the largest territories as predators, moved from Savuti for a while closer to the Khwai Conservancy, Moremi and the Chobe river. Wild dogs can do this because one pack might actually have a territory that covers all these areas. Some packs have been known to have territories that spanned 1,500 sq km, nearly the size of London. Such a vast territory may not hold more than two packs and as the dogs are, by nature, constant wanderers, their moving was their way of adapting to the change.

Lions, too, can have big territories and like any predator, the territory size depends on the availability of food, water and females. The lions of Savuti could not spread too far as different lion prides already had colonised areas everywhere else. Small prey is of no use to the pride. Sometimes, lionesses and lions will go on an individual hunt, but mostly they live and feed as a pride. Small prey will not feed a pride and even a good size buck will feed the pride only for a few days and then they will have to hunt again. So, they learnt to hunt elephant. They avoided the big bulls and adult females, but that still left the young and sub-adults, and even the old or the few isolated members of the herds and the injured.

The cat fish trapped in their mud pools, waiting for the rain waters from Angola to trickle into the Savuti Channel in Botswana.
Photo credit: Craig Foster

One elephant kill meant a lot of meat for many days. Savuti was the first time I saw dozens of dead elephant bones everywhere. Then like a whisper, in 2010, the Savuti Channel filled again, and once more the character of the marsh changed. Now, larger numbers of animals were back and Savuti slowly regained its former glory. The waters stopped once more at the end of 2011 and in 2012 when I reached Savuti, the marshes were dry and the Channel empty. The day I arrived, the first trickle of head waters started in the Channel. Once more, heavy rain in the Angolan highlands had allowed the excess water to run down to the Channel. I got out of the jeep and watched as a small trickle of water, which looked like someone had left the tap on, crept across the sandy bed of the Channel. The thirsty earth seemed to soak it up as it crept forward. I could not understand how this would lead to the Channel being filled at all. It seemed so little.

Further down from this trickle in the dry Channel, another drama was playing itself out. The last time the water had come through, it had brought with it cat fish. They had thrived, grown and spawned. Now those fish were trapped in ever shrinking puddles of mud, their only salvation that small trickle of water kilometers away. The mud puddles were exactly that, brown thick gooey holes filled from edge to edge with fish. They had nowhere to go and would lie still for the most part and then explode into some shaking in order to stir the mud and keep it from hardening and drying up in the harsh sun. The cat fish could only survive if they stayed coated with the wet mud and did not get baked by the sun. The stirring of the mud with their bodies also allowed for more oxygen. I had passed several pools where the fish were already dead, trapped in the mud that had become hard-baked. The sun beat down and I could feel the urgency and the panic.

But in the way that nature has, that trickle became a larger flow, which became a greater stream, which became a babbling rushing brook that started to fill the channel. Walking with the head waters as they flowed and filled was perhaps one of the best nature experiences I have ever had. Life literally flowing with me, changing one thing to another in a blink of an eye. In the morning, I watched dying cat fish trapped with nowhere to go slowly being picked off by marabou storks and leopards. Simply hooked out of their mud puddles and eaten.

Then in the afternoon, I watched the waters come and fill these puddles and fill the channel and I saw the cat fish explode into action, swimming upstream to spawn. Some of them had expended so much energy on simply surviving in the mud puddles that even in the cool waters they just lay there, unable to move, disoriented and floating the wrong way. One big fish almost as long as me and weighing close to 40 pounds just lay on the side gasping and unmoving and I could not bear it. I found myself walking into the water and picking her up. To my surprise, instead of feeling cold, she felt warm and heavy. I think I was just feeling her life essence. She did not move or thrash and lay utterly still. A lifetime of instinct telling her that I was there to help and not a threat. I walked her upstream and finally let her go, urging her to make it.

Feeling that pulse of life in my arms I could not help but think about how life could be abundant and easy and magical, and turn on a cosmic coin toss to become exactly the opposite. The choice is to either knuckle down and fight with all of our energy until the day life becomes easy again, or just die. There is nature, no sentiment, no agenda, simply nature. What makes us think we can destroy that? We can and have ruined her, defiled her, made it harder for her to function, but really all we have done is trapped ourselves in the mud puddle like the cat fish. However, unlike the cat fish we do not have the innate intelligence to let a greater plan for good or bad flow over us, but try and control it and make things worse.

In all the time I spent with uncle Siddharth, learning about nature and how to be in nature, I also learnt one very important thing and that was a reverence for nature. Uncle Siddharth was not a religious man. I am not even sure that he believed in God, or at least not in any conventional way. He was a supremely spiritual man, though, and believed in nature as the singular source. If there was any God he believed in, it was nature. There was immense love and reverence in the way he talked about nature, the way he felt for her. It was the most important thing I ever learnt from him. My father too had the same sentiments and it has been the singular silver thread running right through my life.

As a teenager, I could never have articulated any of this. Even then I took a lot for granted, especially my time with uncle Siddharth, little

Cat fish churning up the mud pools to keep the mud from hardening in the sun.
Photo credit: Craig Foster

knowing that I would lose him a few years after our mad cyclone day at the beach.

Had I known how short our time together would actually be, I would have questioned him more intently about his beliefs, asked him more about how he felt, and spent more time learning from him. That damned permanence of impermanence. Even in that short time I was given with him, it changed my life and gave me the most important and cherished aspect of my life – my bond with the wild world around me. It is perhaps that understanding that will always give me hope even in the midst of what is now the most challenging and sensitive time in our history.

We are living in the future. When I was a child and I used to read about environmental destruction and warnings about what we were doing to our planet, the endgame was always projected well into the future. In the 1980s, dates like 2023, 2030 and so on sounded like very faraway times, times that would take a long time in coming. It felt like we had time in which to right wrongs and fix things. Today, however, as we stand in 2017, 2030 is only as far down the line as 2000 is behind us. Which means the time is going to come and come soon.

It is an indisputable fact that right now we are in the middle of the sixth great mass extinction. The sixth wave of species dieoffs in half a billion years. The rate of dieoff right now, though, is the worst that the world has seen since the disappearance of dinosaurs about 65 million years ago. What makes it worse is that this time this wave of extinction is on us. We did this. Extinction is a natural process in nature. Natural selection makes extinction possible. It is how the system works. However, extinction rates, if natural, are usually at the rate of one to five species per year. Today, we are losing species at the rate of 1,000 to 10,000 times that natural rate per year. Scientific estimates say that we will have lost close to 50 per cent of all species by the year 2050. We stand to lose one-fifth of the world's species by 2025. These are dates that are not so far off.

Ecosystems thrive only because of biodiversity. When species die off, they tend to create a knock-on effect, taking with them other species that are dependent on them, thereby multiplying the rate of loss. This dieoff also allows certain other species to proliferate, causing further damage down the line.

What's worse is that we are not even aware as yet of all the species on the planet. Scientists estimate that so far we have discovered about 1.2 million species; this includes not just mammals, birds, reptiles, insects, amphibians and aquatic species but also plant species. At present, it is believed that we have only discovered about 10 per cent of the species in the oceans. A new study by Canada's Dalhousie University estimates that there are about 8.7 million species on earth. This means we have not even found 86 per cent of the world's species and yet, we are losing them at an alarming rate. There have been disputes over this number but no one disputes the fact that we have only touched the tip of the iceberg when it comes to knowing about all the species on our planet. It is also an undisputed fact that each year, scientists discover about 15,000 new species.

The main reason for this devastating loss is anthropogenic activity, leading to habitat loss.

In India alone, according to the latest data acquired by a group of environmentalists from the Ministry of Environment and Forests through an RTI, about 333 acres of forest land are lost every day. Most of the land is diverted for industrial and development projects. As of September 2011, 47 species of animals were marked critically endangered as listed by the red list of the International Union for the Conservation of Nature, and 49 species have been listed as endangered. Hundreds more are listed as threatened and as vulnerable. Many of the animals on the endangered and critically endangered lists like our big cats and elephants and sharks are vital keystone and apex species of our various ecosystems and their loss will wipe out even that which is not considered especially threatened.

It is estimated that about 21.2 per cent of India is covered by forests. These are figures released by the Ministry for Environment and Forests and Climate Change. Conservationists argue against this figure by stating that the figures include plantations and orchards and groves of eucalyptus, poplars, teak and other plantation species which have questionable conservation values. As of now, less than 5 per cent of these forests are protected. These are the dense rich forests with adequate biodiversity that contain most of India's animal and plant species. It is an unavoidable truth that forests that lie outside the ambit of protected areas are being slowly degraded and that important wilderness areas like scrub jungles, grasslands and wetlands are often overlooked as biodiversity hotspots as they do not fall under the conventional classification of forests. Right now, the Environment Ministry has asked states to identify what they call 'degraded forests' and give up 40 per cent of them to private management. This narrow view that only tree cover makes a forest has been the biggest detriment to India's wild habitats. Also, the tree cover that is planted is usually monoculture of non-indigenous fast-growing species that does not help biodiversity.

And in our budget in 2015, we cut the allocations for environment and wildlife down by nearly a quarter to Rs 1,681 crores. Project Tiger has had its budget cut by 16 per cent. All these cuts while threats from climate change are looming large and the country has already received a warning of what is to come with the bad combination of unchecked progress, lack of environment awareness and unpredictable climates both with the Uttarakhand disaster and the floods in Kashmir. Floods in the North East and Assam has displaced close to 1.5 million people and the disaster is still ongoing. There is no disputing these facts – dams on the upstream rivers and unchecked constructions on

The waters finally reached the mud pool and the fish just exploded into action, swimming upstream.
Photo credit: Craig Foster

flood plains and aquifers and illegal sand mining of river banks and deforestation destabilising river banks, all these have contributed to this disaster. Every year, cyclones ravage our coasts and the allocated budget for coastal management is Rs 100 cores. We have 7,000 kilometres of coastline in India threatened by development and industry and a burgeoning population facing down an inclement sea. Already every year, during and after the monsoons and storm season, thousands of people lose their homes and become displaced with nowhere to go.

And in the middle of these challenges, the government, instead of working on finding solutions, is more interested in extracting greater amounts of coal out of the ground, taking over forest lands for development and making no concessions to cut carbon emissions. Hundreds of more dams are being planned in the North East and the Himalayas while being aware that these are seismically sensitive zones.

One of the lessons I learnt as a child – and it seems astonishing that as adults, we tend to ignore this truth – is an old saying in India *Jungle nadi ki maa hai*. The forests are the mothers of rivers. Close to 70 per cent of India's fresh water is a direct result of our forests.

Basically, our forests help act as catchment areas for the rain. A few months ago, I stood in an old Shola forest high up in the Western Ghats, when suddenly the skies opened up and I could hear the rain pound down on to the canopy. It took a few minutes for the rain to reach me standing under the trees. Even then it was delicate drops of water and not the downpour that I could hear. When it rains, the water is grabbed by the canopy, it then runs down the trunks and seeps into the ground after being absorbed by vines and other smaller plants. In the soil, it is held by the falling leaves and slowly percolates down into the natural aquifers. In other areas, depending on the slope of the land, it accumulates and flows, becoming streams and rivers that then gather into bigger rivers and feed lakes. When trees are cut down, this process is lost. The water just pours down instead of being gentled and the force of it washes away the top soil which is invariably eroded by the wind when there are no plants and roots to hold it in place. This prevents the water from being sequestered as it should and also causes massive amounts of soil to mix with the flowing water, creating enormous amounts of silt which can block and change the flow of the streams and rivers. The evaporation of the water from the soil is also very fast when there are no plants and trees and leaves on the ground to cover the soil and maintain its moisture.

Now, there is an even more fascinating theory of the biotic pump, which looks at the fact that it might not be the rains that create rainforests or even lush forests but the forests that might be creating the microclimate that creates the rain. This theory has been proposed by Russian researchers Anastassia Makarieva and Victor Gorshkov. Forests, it is being theorised, pull in large amounts of water vapour from surrounding regions and from nearby bodies of water if there are any to be found. This vapour condenses into rain. When there is a lot of rain, the local atmospheric pressure can drop which then creates a kind of vacuum that sucks in more water vapour from outside the forest which then repeats this whole process again. If this theory holds true, then there is greater potential danger to our cutting down forests, like the Amazon for instance.

In India, our rainforests occur in the Western Ghats, the North East and the Andaman and Nicobar islands. Both the Western Ghats and the forests of the North East are under huge pressure for development. In

The cat fish have been trapped in these shrinking pools for months and the waters are their last hope for survival.
Photo credit: Craig Foster

the Western Ghats there were also extensive grasslands. The indigenous people of the area revered these grasslands but modern India did not. Across the country, grasslands are seen as 'wasteland' and are given over for development or cultivation or even plantations. A grassland acts as a sponge and holds water through the year, even in the lean season. Streams that percolate and collect and flow through them stay steady through the year. Plantations of exotic trees, the invasion of exotic weeds and, of course, development projects stop this from happening. Wetlands across the country too have been converted into agricultural fields and many have been drained to reclaim land for construction. Wetlands are the bastion that stand against a raging river and the rest of the landscape. Wetlands absorb the excess of flooding and break the damage a river in spate can cause. They also allow for the excess water to be stored creating an invaluable ecosystem of their own one that is filled with birds, fish, shrimp, snails, otters and mongoose and so much more.

Recently, in a bid to open up the Western Ghats for development projects, the government appointed a committee to look into the matter. The Western Ghats ecology expert panel, also called the Gadgil Commission, named after its chairman Madhav Gadgil, was an environmental research commission appointed by the Ministry of Environment, Forests and Climate Change in 2011. The expert panel after a year of study of the ecology of the region and after consulting with conservationists, specialists, experts and civil society groups, recommended that about 67 per cent of the Western Ghats be designated as ecologically sensitive and kept entirely out of the purview of development. The Western Ghats have been designated as a world heritage site by UNESCO. It is a unique ecosystem with endemic species found nowhere else in the world. It is also the heart of the catchment area for most of the rivers in the water-starved South of the country. The Western Ghats spread out over six states – Kerala, Maharashtra, Tamil Nadu, Goa, Gujarat and Karnataka. The Congress government of the time, dismayed by this, set up a second committee for another analysis. This report called the Kasturi Rangan Report was more favourable towards development, stating that only 37 per cent of the Western Ghats were ecologically sensitive. Of course, the government leapt on to this report.

In Africa, too, the race for development, the civil wars created by the greed to exploit the natural resources, the complete disregard of indigenous knowledge and systems and a growing population is ruining habitats and causing species dieoffs.

It is indeed ironic then that in a future that is bearing down upon us, the two places that will be worst affected by the disasters of a changing climate, rising sea levels and temperatures, droughts, floods and disease will be India and Africa. One, because of such a huge population density of close to 400 people per sq km and the other, as a continent that does not have the financial resources with which to cope. One is a country, the other a continent, but both are mega-diversity areas. It is that which makes both places so incredibly important to the world.

India can stand as an example to the world. Show the world that large numbers of people and wilderness can co-exist. It is a gift that we still have mega fauna that walk our lands amidst all the people and the noise. We just need to be proactive and encourage this centuries old traditional love and tolerance of nature and understand that flamingos and leopards are what make Mumbai city unique for example, and not a sea link road or a high rise. It is the fact that we can boast of the

The trapped fish were a feast for other animals. Here, a marabou stork is catching them.
Photo credit: Craig Foster

lion, various leopard species and tiger – no other country in the world can boast of that – that should make us proud. Not highways that cut through forests and coal mines. We are one of the 17 mega-diverse countries of the word. We hold three of the world's 34 biodiversity hot spots. India is home to 7.6 per cent of the world's mammals, 12.6 per cent of the world's birds, 6.2 per cent of the reptiles and 6 per cent of the flowering plant species. 33 per cent of our plants are endemic, which means they are found nowhere else. We are also home to 17.5 per cent of the world's humans. We are right now the second most populated country in the world and are set to overtake China by 2022, again a date not so far into the future.

Africa is the mother continent. It is where we all came from. Every human on earth can trace her/his genetic lineage to Africa. Africa also holds the oldest repository on human-wildlife interactions. Traditional knowledge and ways of living that can hold solutions to problems we face today. In many countries in Africa too, people and wildlife live side by side and struggle to survive together.

And the bad news does not end there. Australian microbiologist Dr Frank Fenner has predicted that humans will face extinction by the year 2100 due to overcrowding, declining resources and climate change. That is just three, maybe four generations down the road. This is the biggest tragedy to how we are dealing with the earth because we are, with every step, dooming future generations. Most of us in my age group at least have children. We love our children and we want the best for them. How is it that the idea of clean air and water is not number one on our list of what we want for our children? How is it that we can live in a time of perhaps the greatest access to information and knowledge and yet function like ignorant idiots?

What are we teaching our children? It is a fact that children today can identify more brands, cars and movies than they can the animals in their garden. If they have a garden. If we ourselves, as adults, care more for lifestyle than quality of life, how can our children learn any better?

When I was a child, I was lucky that I had an uncle Siddharth and my father.

The question is, are we prepared to look our grandchildren in the eye and say, we knew better but were too selfish to change the way we did things, so, sorry that your world now sucks?

THE PERMANENCE OF IMPERMANENCE

But the beauty is it does not have to continue this way, it can change. In the 5th century, most of the western world thought that the earth was flat and until the 16th century and Copernicus, that the sun revolved around us. People were burnt at the stake for saying differently. But that changed. In our lifetime, we have seen the Berlin wall fall, the Soviet Union emerge from Communism, South Africa's shocking apartheid system dismantled. So many things that seemed like they would never change changed because people put their mind and hearts together and did so. India herself threw of the yoke of British imperialism and became free.

While we are staring down at the mass extinctions, we have also fought and helped take several species that were on the brink of extinction off the list. Today, on an overall average, deforestation has actually reduced; it's just that there are far less forests than there used to be, so the rate of deforestation is still high. The point is we have managed to bring about change when we needed to. And now is the time when we need to, more than ever. This next decade will decide the fate of our world. Whether ecocide or ecosense will prevail is in our hands.

After uncle Siddharth died, my father and I could not bring ourselves to go to the Theosophical Society and walk. We could not face the beach or the broken bridge. A month went by and our weekends were spent doing everything except what we once did. Then one weekend, my father decided that we must go back. So reluctantly, I went with him. We walked the familiar paths, I hugged my trees. I watched barbets, treepies and mynahs and bulbuls and warblers and parrots and squirrels and the odd jackal all run around like nothing has changed. It brought me a strange kind of peace.

We walked towards the setting sun and sat on the broken bridge looking out to sea. This was our way of saying goodbye, but that we will never forget. As we settled in, a most wonderful thing happened. A beautiful brahmini kite circled down towards us and came to a graceful rest on the bridge an arm's distance from me. The bird that had been the first wild experience I had had with my uncle Siddharth. The bird, that had screamed and dived into the ocean on my first ever walk in the Theosophical Society. The bird, that first started me on this path of trying to understand the natural world around us. His snowy white

Maribou stork with fish in its beak.
Photo credit: Craig Foster

head glowing in the setting sun. I felt goose bumps run up my body and my mouth felt dry. There was a strange feeling of static electricity in the air. To me, it was as if uncle Siddharth had come to sit next to me like he had a thousand times before. His own snowy white hair glowing in the setting sun. My father and I watched the sun sink and the bird stayed with us. Finally, as the gloaming occurred, that moment between bright and dark, the bird took off with its distinctive cry and all I heard in the brush of air against my cheek was uncle Siddharth saying, *the permanence of impermanence, my girl.*

There are more things on heaven and earth as the Bard has said. That has been my driving force and my hope. The permanence of impermanence. The resilience and wonder of nature.

It is why even when the challenges look insurmountable and bleak, I feel confident that, as humans, we have the ability to change things for the better.

I hope it happens soon.

The most important landscape for wildlife is the human heart.
                                                    – Bittu Sahgal

AN INTERVIEW WITH DAVID ATTENBOROUGH

If we damage the natural world in the end we damage ourselves.
– David Attenborough

IN 2015, on a wish and a prayer, I called a number I had acquired for Sir David Attenborough, the legendary naturalist and television presenter. I was told the number belonged to his daughter who arranged all his appointments. I dialed, hoping to be able to convince her to let me do a sit-down interview with Sir David. At eighty nine, he was not doing too many long sit-down interviews anymore. The last person with whom he did a sit-down interview was President Obama. I was prepared to either be denied or be told that he may give us a few minutes. I dialed the number. The phone was picked up by none other than Sir David and through my shocked stutterings and stammerings he agreed to give me a long interview. I was in Cape Town at that point and I dashed to Delhi to get my visa and then dashed to London to make the greatest appointment of my life. I had interviewed him once before in 2006, for all of ten minutes. This time, too, he was unbelievably gracious, kind, charming and inspiring. He even helped us carry our equipment out to the car in the pouring rain.

Q: *Sir David, you have now spent a lifetime doing what you do. What sparked your interest?*

A: It was an interest in the natural world. I don't think that a child has been born yet who does not have an interest in the natural world. Of course some people may lose it because there are lots of other attractions as one grows up but if you lose the interest in the natural world you have lost a great treasure, a great source of joy, a great source of solace. It is one of the most valuable things we have.

Q: *Sir David when you joined the BBC in 1952, you were a producer on The Talk. You were not on air. In fact, a senior executive said you did not have the presence to be on air. But here we are, so how did that come about?*

A: I had no intention to appear on air. I had a degree in natural sciences and I had served in the Navy and I was working in a publishing house. It was all extremely boring. I simply answered a request from BBC radio and I didn't get an interview but then I got a letter saying that we have a sort of new thing called television and right now no one is very interested in it but we think you could have a future in it, so would you like to come do that? It's extraordinary when you see that today how desperately people want to get into television and at that time people were saying please come and try it. I went there as a trainee and then as a producer and I produced programmes of all kinds. Quiz programmes, garden programmes and one or two animal programmes. We were going so fast it was very easy to specialize very quickly so I was able to move into the area which fascinates viewers, which is natural history.

Q: *So you started with Zoo Quest and then around 1979 you did the Life on Earth series and from then you have done a range of Life series. You have travelled across the world and have gone literally from the black and white television era all the way to 3D television. If I were to ask you to pick your top three moments in the natural world which ones would they be?*

A: Well, one of them would most certainly be when I dived on to a coral reef. The first time you dive with aqualung gear and become independent of gravity is very exciting. The thought of being free – where with a flip of your flipper you can move up or down or sideways and are free of the bonds of gravity – is amazing. But when you look down and see a coral reef which is a universe of colours and shapes and creatures you never even knew existed, creatures of stunning beauty, it certainly is one of the great moments. There have been lots of others, of course, but that was a transformational one because you really move into another world.

The other great memories have been with apes of one sort or another. I mean you can look in the eyes of a tiger but you don't know what it is like to be a tiger but when you look in the eyes of an ape you can imagine what it is like to be an ape. There is a recognition with apes that you get, I suppose some people get it with dogs. For me, it is apes really.

Q: *You have literally travelled from pole to pole and it has been an enormous career but doesn't it disillusion you at times that we are at this point right now where things seem worse than ever?*

A: Yes, now they are worse than ever. That is very true. But on the other hand, they could have been a good deal worse if conservationists from all around the world had done nothing and were not doing something to try and slow down the damage. So the fact that it is depressing should be reason for renewal of action, not reversal.

Q: *And in December 2015, in Paris the UN is going to meet on climate change and this is COP 21 which clearly indicates that the previous 20 COPS haven't gone very far. You have called climate change the greatest threat and you have done a show, The State of the Planet, on it. Why do you think it is so hard for us to do something about this or get people to actually put their money and minds behind it?*

A: It has never happened in history has it? I mean it has never ever, ever happened in the history of the planet that all nations of all kinds and conditions and race have all got together and agree to do something. So, the fact that it is has not happened before is no reason not to try to get it done now. And the reason for doing it now is the action taken by one nation which used to be more or less restricted to what they did with their own frontiers is now world wide. If you poison one side of the Pacific in North America it is felt on the other side of the Pacific in Asia. So there is every reason why nations should try to get together and settle their differences and work on an uniform plan to deal with the danger which is a severe threat to every man, woman and child alive.

Q: *You have often stressed that one of the things we don't bring into our discussions on biological collapse or climate change is population numbers. To quote a statement you made in Life of Mammals: 'Perhaps the time has now come to put that process into reverse, that instead of controlling the environment for the benefit of the population perhaps it's time we control the population for the benefit of the environment.' When you started out in 1950 there were about 2.5 billion on the planet and today we are at 7 billion, set to grow to 10 billion. Why are we not talking about this?*

A: Because giving birth and producing a child is one of the most precious rights and freedoms for a human being and who am I to tell somebody what they should do on that issue? It's a very precious issue but nonetheless we can't avoid it and we have to grapple with it. One consolation I have, one source of hope, is recognition that wherever woman have the political right and the moral right and the medical facilities and the education and literacy to control their own bodies, the birth rate falls.

Q: *Many people still deny climate change and say the earth has been shaped by violent forces at various different parts of time. They say there have been huge climatic or natural events like massive volcanoes so the fact that the we are now looking at climate change is actually anthropogenic pressure that's creating all these problems or adding to the existing problem. So what would you have to say to people who still deny climate change?*

A: Well there is a reason to deny climate change. It is so much easier to say it is not happening and we can turn our eyes away from it and not recognize it and the money required. It costs money to deal with climate change, it doesn't make money and so there is every reason why a lot of people turn away from climate change. But if you do recognize the fact and you care about your children and your grandchildren and what world you are going to leave them then you have to tackle that problem and deal with people who deny climate change. Progress has been made, you know. I mean American President Obama knows very well about climate change but they live in a democracy in America and you have to get the democratic process working and you have to get the people on board. Progress has to be made so we can all live. It would have been absurd to suppose people could get a few nations together and they would deal with the problem. The nations are coming together now. The fact they have failed so far in previous conferences is no reason why we shouldn't keep on trying. We have to keep on trying.

Q. *One of the things about the whole climate change debate is we focus on cutting down emissions and it's all about fossil fuels and their use. That's where the focus really is – on economics and climate change.*

A: You are quite right. The problem with dealing with climate change is until now it has always been don't do this, stop doing that. It's always been prohibition. It is always being costing money and this time there is a new plan, which aims to get an inclusive report. If we could produce energy from renewable resources, non - damaging resources at a cheaper price than fossil fuels the problem would disappear and in fact the world would be wealthier because we won't spend the money on these very difficult problems of getting more energy from fossil fuels. And that is possible if nations of the world would collaborate scientifically to solve problems of, first of all, generation from renewable resources, of transmission of resources and storing of energy. If you get those at a cheaper price you would solve the problem absolutely. If the Americans can put a man on the moon in 10 years, are you telling me that the coordination of the scientific brains of the whole world put together on a proper platform couldn't solve what is relatively simple. I mean the basic technology is there. We know it can be done. What we have to do is improve it to make it more efficient and cheaper.

Q: *Sir David in your extensive travels the one country that you haven't travelled that much in is India. Is there a reason for that?*

A: I know that there are very good Indian filmmakers so why do they need me? I have travelled through Rajasthan and further south and had a wonderful time but Indian filmmakers are very skilled and good in their job and they don't need me. That doesn't mean that we have not shown Indian natural history on our programmes. We have.

Q: *And speaking of Rajasthan I think one of your not so favorite animals would be the rat? You went to the rat temple in Bikaner, didn't you?*

A: Yes. I have to confess. Well rats carry disease and rats have no respect for human beings. So think about it, most wild animals if you don't want to get close to them you don't have to. They do not come up to you that much. They say that you are never ever far away from a rat no matter where you are. Yes I was in the rat temple. There were rats running on my feet. I did not enjoy it. But maybe that is my problem, my limitation, and I should be more universal in my loving.

Q: *Several new species and even a fossil has been named for you, like the Attenborosaurus Conybeari.*

A: Yes, That's nice.

Q: *Yeah! When you started collecting fossils as a young boy did you see your life go in this direction?*

A: I couldn't. It's a funny business. I mean it's the greatest compliment that a naturalist can give another naturalist. I am very honored that it should be so. I mean the Attenborosaurus Conybeari which you mentioned is one of the more prominent ones. It's a fossil of the great sea reptile that we found in England, which was discovered in the 19th century. But all this depends on where the paleontology goes. It occasionally revives the groups and you realise that some aren't what we thought and that they didn't belong in this family and that they belonged to a completely new family so they invented this new family and gave it my name.

Q: *You are seen as a national treasure and you have been in many, many polls that voted you as one of the most trusted people on the planet. Your voice is pretty much the most recognizable one on TV. Most people can recognize it. Do you sometimes feel that if you stood up right now and said the moon was made of green cheese, about a million people would believe you so. Does that put pressure on you because you have that ability to reach so many people with all these positive messages?*

A: It makes one think that one should be careful about what one says and I suppose I can. My life has been spent as a public service spokesperson and I started as a producer and, as a producer, I knew very well that my job was to produce programmes that weren't propaganda and weren't granting access to biases and that took a fairly objective view. I therefore knew that I have to be very careful. It's a perfect truth that while privately I was convinced that global warming was happening, climate change is happening, I didn't make programmes about it until I was absolutely sure that the evidence was conclusive. I was not showing that it can be an opinion. I was actually conveying an accepted fact that was accepted by the vast majority of scientists involved in that issue.

Q: *Speaking of that, you are primarily a scientist, it's paleontology, you have studied natural history. But when I watch you in your programmes, while there is of course that part of it, I feel there is an overall reverence that comes through in the way you enjoy being in the natural world.*

A: Well, I am pleased that you should say that that's the case. I do think the natural world is a treasure and I do think that human beings do not have a right to simply dispose of it in any way they wish. I think that the natural world has to be respected. It is one of the greatest treasures that humanity has. The fact is that we are a part of the natural world and we depend on it. Every mouthful of food we eat depends on the natural world, every breath of air depends on the natural world and we are a part of it. If we damage the natural world, in the end we damage ourselves. We are putting enormously more pressure on the natural world as the human species than any other single species until now. We have to recognize our responsibilities.

Q: *Sir David, you have signed papers against creationism in education curriculums and you have, I think, called yourself an agnostic at a level and have said the truth you see is through evolution really. But there is something to be said for not a conventional religion, but a certain reverence or a spiritual feeling towards the natural world. If you take a country like India with 400 people per square kilometer, we still have the tiger, the lion, the leopard. We have mega fauna. We are one of the mega-diverse countries in the world and that's because of a certain spiritual reverence that people have for the natural world. So, where do you stand on that?*

A: I would like to think that I revere the natural world and that I recognise its rights and its mysteries. We don't understand what life is and if you ask me about my religious faith I will say that I am agnostic, as you said, but at the same time, I think of myself cutting a hole in a termite home and looking, I am seeing all these little individuals busy with their jobs. Now, they will never recognise me because they don't have eyes that see that way. They are blind, so they can't tell whether there is something bigger, that there is some bigger world outside. They don't know. Well, maybe there is something that we don't know or a few very select individuals have an insight about this. I believe that is the case. I haven't found the answers to all my questions but then why would there be something like answers to all questions? After all it is a mystery and that's why humans are absorbed in it.

Q: *What gives you your energy to go on? You once said 'If I were earning my money by mining coal, I would be very glad indeed to stop but I am not, I am swirling around the world looking at the most fabulously interesting things. Such good fortune'. Is that what you are going to continue doing? Swirling around the world?*

A: That seems be a very good answer to me. Yes, I mean I also want to say that I am blessed and that in itself is a sort of a curious statement. So I wouldn't say that. What I would say is this, that I am just coming up to my 90th birthday and I think there are an awful lot of people in the world who are not as lucky as me simply in terms of health. I can still walk about. I can still walk distances. I can still put two words together and it seems to me almost a crime not to continue. The second part of that is because so many people would give so much to have the sort of situation that I am in, so I feel it would be stupid not to take advantage of it.

Q: *Isn't it ironic at a level that we live in a time of great knowledge or access to knowledge and yet we live like we know nothing?*

A: I am not sure whether we know nothing or whether we understand everything is the problem. Understanding and sympathy, I mean you can go on to know all kinds of facts. I am surrounded by facts on bits of paper and I don't know them. But I would like to think that I understood a little, not much, but a little. And so that's why the process of finding out in itself is important. So, not knowing, it's not virtue, it's called enjoying yourself.

Q: *My last question. If the delegates at the Paris conference asked you what were three important things they should pay attention to what would you say to them?*

A: Well, the first would be that scientists should collaborate to solve these problems about renewable energy. However at the back of my mind I have a little concern about whether human beings – if they have that energy – will they use it well or will they use it to destroy more of the natural world. I mean would it be like say, hurray we can now cut even more trees even more cheaply? So the second thing would be an understanding of the effects of what we are doing if you get the energy. I suppose the third thing would be that we could recognize we live in a finite world and therefore we cannot keep growing infinitely. The moment has to come when the earth cannot produce enough food in the extra evidently luxurious way we use the planet's resources and there will be starvation and worse, if we allow the human population to grow and grow and grow.

Q: *Sir David, thank you so much for joining us. In India when a person turns 80 we say they have seen a thousand full moons. I wish you a thousand more full moons because we all – not just the natural world – need you to keep swirling about and inspiring us with miracles.*

A: Thank you. That was a very gracious interview.

After we had restored his space to how it looked before we came in with all our equipment and lights, he took me on a small tour of some of the fossils he had collected over his years of travel. He quizzed me on them and I got all the answers right. Sheer luck on my part as I was familiar with some of them from my readings and some from the fact that my husband is also equally obsessed with fossils and bones. However that pleased him enough to invite me back to visit him whenever I made another trip to England.

He is 90 years old and in that hour he gave me more energy than I had felt in years. Filled me with more optimism and energy than I had been feeling. It gave me a renewed sense of the fact that each of us can only do what we can do and whatever that is, however little is still better than doing nothing. I walked away from that interview feeling equally blessed about my life.